T0140910

Physiological responses of aquatic macrophytes to natural organic matter: *potential for structuring aquatic ecosystems*

Dissertation

zur Erlangung des akademischen Grades

doctor rerum naturalium
(Dr. rer. nat.)
im Fach Biologie - Ökologie

eingereicht an der
Mathematisch-Naturwissenschaftlichen Fakultät I
der Humboldt-Universität zu Berlin

von
Sheku Kamara
geboren am 20.02.1970 in Timbo village (Tonko-Limba), Sierra Leone

Präsident der Humboldt-Universität zu Berlin
Prof. Dr. Christoph Markschies

Dekan der Mathematisch-Naturwissenschaftlichen Fakultät I
Prof. Dr. Lutz-Helmut Schön

Gutachter/innen:
1. PD Dr. Stephan Pflugmacher
2. Prof. Dr. Werner Kloas
3. Prof. Dr. Mark O. Gessner

Tag der mündlichen Prüfung: 10.12.2008

Gedruckt mit Unterstützung des Deutschen Akademischen Austauschdienstes

Bibliografische Information der Deutschen Nationalbibliothek

Die Deutsche Nationalbibliothek verzeichnet diese Publikation in der Deutschen Nationalbibliografie; detaillierte bibliografische Daten sind im Internet über http://dnb.d-nb.de abrufbar.

ISBN 978-3-8325-2115-8

Logos Verlag Berlin GmbH
Comeniushof, Gubener Str. 47,
10243 Berlin
Tel.: +49 030 42 85 10 90
Fax: +49 030 42 85 10 92
INTERNET: http://www.logos-verlag.de

Table of Contents

Summary I

Introduction 1

1.1 Background (Problem statement) 1

1.2 Sources of natural organic matter (NOM) 2

1.2.1 *Quercus robur* (oak) leaf litter 4

1.2.2 *Phragmites australis* (reed) leaf litter 5

1.3 Reactive oxygen species (ROS) formation 5

1.4 Oxidative stress and ROS signalling 6

1.5 Mechanism of defense against oxidative stress in plants 7

1.5.1 Enzymatic defenses 8

1.5.1.1 Superoxide dismutases (SOD: EC 1.15.1.1) 9

1.5.1.2 Peroxidases (POD: EC 1.11.1.7) 9

1.5.1.3 Catalases (CAT: EC 1.11.1.6) 9

1.5.1.4 Glutathione reductase (GR: EC 1.8.1.7) and glutathione peroxidase (GPx: EC 1.11.1.9) 10

1.5.1.5 Glutathione S-transferases (GST: EC 2.5.1.18) 10

1.5.2 Non-enzymatic antioxidant defenses 10

1.6 Lipid peroxidation 11

1.7 Gene expression 12

1.8 Photosynthesis and oxidative stress 12

1.9 Aquatic macrophytes 13

1.10 Acclimation 14

1.11 Aim of study 14

2. Materials and methods 16

2.1 Materials 16

2.1.1 Equipments 16

2.1.2 Chemicals 16

2.1.3 Plant materials 18

2.2 Methods 18

2.2.1 Preparation and chemical characterisation of leaf extracts ... 18
2.2.2 Exposure to leaf-litter extracts ... 19
2.2.2.1 Dose-response exposures ... 19
2.2.2.2 Time-response exposures ... 20
2.2.3 *Lemna minor* bioassay experiments ... 21
2.2.3.1 Determination of *Lemna* growth rate and inhibition ... 23
2.2.4 Measurement of photosynthesis ... 23
2.2.5 Pigment analysis ... 24
2.2.6 Antioxidant (GSH and GSSG) determination ... 24
2.2.7 Intracellular hydrogen peroxide determination ... 25
2.2.8 Multi-enzyme preparation ... 26
2.2.9 Determination of protein concentration ... 27
2.2.10 Biochemical assays ... 28
2.2.10.1 Superoxide dismutase activity assay ... 28
2.2.10.2 Catalase activity assay ... 29
2.2.10.3 Peroxidase activity assay ... 29
2.2.10.4 Gutathione reductase (GR) activity assay ... 30
2.2.10.5 Glutathione peroxidase (GPx) activity assay ... 31
2.2.11 Measurement of lipid peroxidation ... 34
2.2.11.1 Calculation of Lipid peroxidation ... 34
2.2.12 Molecular biology experiment (GR gene isolation and expression) ... 35
2.2.12.1 RNA isolation ... 35
2.2.12.2 Reverse transcription ... 36
2.2.12.3 Semi-quantitative RT-PCR ... 36
2.2.12.4 Primer design ... 37
2.2.12.5 Purification of DNA from gel and sequencing ... 37
2.2.13 Statistical Analysis ... 38

3. Results ... 40

3.1 Experiments with *Ceratophyllum demersum* ... 40
3.1.1 Experiment 1: Dose-response effects of *Phragmites australis* and *Quercus robur* leaf litter extracts on *C. demersum* ... 40
3.1.1.1 Dissolved organic carbon fractions in *P.australis* and *Q. robur* leaf extracts ... 40
3.1.1.2 Photosynthetic oxygen release in *C. demersum* ... 40
3.1.1.3 Chlorophyll pigments in *C. demersum* ... 41
3.1.1.4 Enzymatic activity response of microsomal glutathione S-transferase (mGST), assayed with the conjugating substrate 1-chloro-2,4-dintrobenzene (CDNB), in *C. demersum* .. 43

3.1.1.5 Enzymatic activity response of cytosolic glutathione S-transferase (cGST), assayed with the conjugating substrate CDNB, in *C. demersum* 43
3.1.1.6 Enzymatic activity response of cytosolic glutathione S-transferase (cGST), assayed with 4-hydroxynonenal as conjugating substrate, in *C. demersum* 44
3.1.1.7 Enzymatic activity response of glutathione reductase (GR) in *C. demersum* 45
3.1.1.8 Enzymatic activity response of guaiacol peroxidase (POD) in *C. demersum* 45
3.1.1.9 Lipid peroxidation (LPO) in the aquatic macrophyte *C. demersum* 46
3.1.2 Experiment 2: Acclimation experiments with *C. demersum* exposed to *P. australis* and *Q. robur* extracts 47
3.1.2.1 Photosynthetic response on time-dependent exposure of *C. demersum* to *P. australis* and *Q. robur* extracts 47
3.1.2.2 Enzymatic response of microsomal glutathione S-transferase (mGST), measured with CDNB, on time-dependent exposure of *C. demersum* to *P. australis* and *Q. robur* extracts 48
3.1.2.3 Enzymatic response of cytosolic glutathione S-transferase (cGST), measured with CDNB, on time-dependent exposure of *C. demersum* to *P. australis* and *Q. robur* extracts 48
3.1.2.4 Enzymatic response of cytosolic glutathione S-transferase (cGST), measured with 4-hydroxynonenal (4-HNE), on time-dependent exposure of *C. demersum* to *P. australis* and *Q. robur* extracts 49
3.1.2.5 Enzymatic response of guaiacol peroxidase (POD) on time-dependent exposure of *C. demersum* to *P. australis* and *Q. robur* extracts 50
3.1.2.6 Enzymatic response of glutathione peroxidase (GPx) on time-dependent exposure of *C. demersum* to *P. australis* and *Q. robur* extracts 51
3.1.2.7 Enzymatic response of glutathione reductase (GR) on time-dependent exposure of *C. demersum* to *P. australis* and *Q. robur* extracts 52
3.1.3 Experiment 3: Effects of *P. australis* and *Q. robur* leaf litter extracts on intracellular hydrogen peroxide content, glutathione content and glutathione reductase gene expression in *C. demersum*. 53
3.1.3.1 Hydrogen peroxide content in the aquatic macrophyte *C. demersum* 53
3.1.3.2 Total glutathione contents in the aquatic macrophyte *C. demersum* 54
3.1.3.3 Reduced to oxidised (GSH/GSSG) glutathione ratio in the aquatic macrophyte *C. demersum* 55
3.1.3.4 Glutathione reductase expression in the aquatic macrophyte *C. demersum* 56

3.2 Experiments with *Lemna minor* 59
3.2.1 Experiment 4: Dose-response effects of *P. australis* and *Q. robur* leaf extracts on *L. minor* 59
3.2.1.1 Intracellular hydrogen peroxide content in the aquatic macrophyte *L. minor* 59

3.2.1.2 Total glutathione content in the floating aquatic macrophyte *L. minor* 60
3.2.1.3 Reduced to oxidised (GSH/GSSG) glutathione ratio in the floating aquatic macrophyte *L. minor* 60
3.2.1.4 Enzymatic activity response of superoxide dismutase in the floating aquatic macrophyte *L. minor* 61
3.2.1.5 Enzymatic activity response of guaiacol peroxidase in *L. minor* 62
3.2.1.6 Enzymatic activity response of catalase in the floating aquatic macrophyte *L. minor*... 62
3.2.1.7 Enzymatic activity response of microsomal glutathione S-transferase (mGST), assayed with the conjugating substrate 1-chloro-2,4-dinitrobenzene (CDNB), in *L. minor*........ 63
3.2.1.8 Enzymatic activity response of cytosolic glutathione S-transferase (cGST), assayed with the conjugating substrate 1-chloro-2,4-dinitrobenzene (CDNB), in *L. minor*. 64
3.2.1.9 Enzymatic activity response of cytosolic glutathione S-transferase (cGST), assayed with the conjugating substrate 4-hydroxynonenal (4-HNE), in *L. minor* 65
3.2.1.10 Enzymatic activity response of glutathione peroxidase (GPx) in the floating aquatic macrophyte *L. minor* 65
3.2.1.11 Enzymatic response of glutathione reductase in the aquatic macrophyte *L. minor*...... 66
3.2.1.12 Lipid peroxidation in the floating aquatic macrophyte *L. minor* 67
3.2.2 Experiment 5: Acclimation experiments with *L. minor* exposed to *P. australis* and *Q. robur* extracts 68
3.2.2.1 Enzymatic response of guaiacol peroxidase (POD) on time-dependent exposure of *L. minor* to *P. australis* and *Q. robur* extracts 68
3.2.2.2 Enzymatic response of catalase on time-dependent exposure of *L. minor* to *P. australis* and *Q. robur* extracts 68
3.2.2.3 Enzymatic response of microsomal glutathione S-transferase (mGST), measured with 1-chloro-2,4-dinitrobenzene (CDNB), on time-dependent exposure of *L. minor* to *P. australis* and *Q. robur* extracts 69
3.2.2.4 Enzymatic response of cytosolic glutathione S-transferase (cGST), measured with CDNB, on time-dependent exposure of *L. minor* to *P. australis* and *Q. robur* extracts 70
3.2.2.4 Enzymatic response of cytosolic glutathione S-transferase (cGST), measured with 4-hydroxynonenal, on time-dependent exposure of *L. minor* to *P. australis* and *Q. robur* extracts 71
3.2.2.5 Enzymatic response of glutathione peroxidase on time-dependent exposure of *L. minor* to *P. australis* and *Q. robur* extracts 72
3.2.2.5 Enzymatic response of glutathione reductase on time-dependent exposure of *L. minor* to *P. australis* and *Q. robur* extracts 73
3.2.3 Experiment 6: *L. minor* bioassay experiments 73
3.2.3.1 Chlorophyll pigments in *L. minor* exposed to *P. australis* and *Q. robur* extracts 73

3.2.3.2 Photosynthesis in *L. minor* exposed to *P. australis* and *Q. robur* extracts 75

3.2.3.3 Number of fronds in *L. minor* exposed to *P. australis* and *Q. robur* extracts 76

3.2.3.4 Growth rates in *L. minor* exposed to *P. australis* and *Q. robur* extracts 77

3.2.3.5 Toxicity dose-response analyses in *L. minor* exposed to *P. australis* and *Q. robur* extracts 78

4. Discussion 86

4.1 Discussion: Experiments with Ceratophyllum demersum 86

4.1.1 Physio-biochemical response experiments 86

4.1.1.1 Experiment 1 (Short-term 24 h dose-dependent): Response of the photosynthetic and antioxidative system of *C. demersum* to leaf litter-derived DOC 86

4.1.1.2 Experiment 2: Acclimation (time response) of *Ceratophyllum demersum* to the effects of leaf litter-derived DOC 89

4.1.2 Molecular and antioxidant response experiments 93

4.1.2.1 Experiment 3: Regulation of glutathione reductase gene expression and glutathione redox dynamics in *C. demersum* by leaf litter-derived DOC 93

4.2 Discussion: Experiments with Lemna minor 96

4.2.1 Biochemical responses of *L. minor* to leaf litter-derived DOC 96

4.2.1.1 Experiment 4: Short-term (24 h) biochemical dose-responses 96

4.2.1.2 Experiment 5: Acclimatory (time) responses of *L. minor* to the effects of leaf litter-derived DOC 97

4.2.2 *Lemna minor* Bioassay Experiments 98

4.2.2.1 Experiment 6: Photosynthesis and growth responses of *L. minor* L. 98

4.3 Discussion: Overview of the physiological responses of *C. demersum* and *L. minor* to leaf litter-derived DOC 100

5. Conclusions and perspectives 101

6. References 104

List of abbreviations 114

Declaration 116

Publications and Conference participation 117

Acknowledgements 119

Curriculum Vitae 120

Summary

Dissolved organic carbon (DOC) levels in aquatic systems are influenced by various factors ranging from anthropogenic perturbations (e.g. land-use systems) to natural processes related to weather phenomena. Natural processes include the large amount of leaves shed by deciduous trees every year, which degrade and release chemicals into the aquatic system. It is estimated that up to 30% of the total DOC in streams is contributed by terrestrial leaf litter alone. Although leaf litter decomposition rate as influenced by various ecological factors and the contribution to carbon flux in the aquatic system has been aptly investigated, the potential implication for aquatic biota of the decomposition products has not received parallel attention.

In the present study, leaf litter-derived DOC from terrestrial and aquatic plants were used to study the impact of leaf-litter decomposition products on the aquatic macrophytes *Ceratophyllum demersum* and *Lemna minor*. Photosynthetic oxygen release, marker enzymes of the antioxidative system, glutathione redox dynamics and growth rate were used as testing parameters. The overall objective was to evaluate the potential of leaf litter-derived DOC to interact and influence aquatic macrophytes by imposing oxidative stress leading to cellular damage and photosynthetic and growth reduction, thereby manifesting a functional link between terrestrial and aquatic ecosystems. The potential for the studied macrophytes to acclimate to the effects during prolonged (7 days) exposure was also evaluated

For this purpose, aqueous oak (*Quercus robur*) and reed (*Phragmites australis*) leaf litter extracts were used as allochthonous and autochthonous sources of DOC, respectively. Dose- and time-dependent exposure experiments were carried out and various biochemical (enzymatic and non-enzymatic antioxidants), physiological (photosynthesis and growth) and molecular (glutathione reductase gene {GR} expression) endpoints were assessed.

Significant photosynthetic inhibition was observed at low and high DOC levels in oak and reed extracts. In *C. demersum*, photosynthetic reduction could not be explained by chlorophyll pigments since they were not adversely affected. In *L. minor* on the other hand, chlorophyll pigment reduction accounted, at least in part, for photosynthetic inhibition, accompanied by growth reduction at high DOC levels.

Biotransformation and antioxidative enzymes were induced at DOC levels ranging from 0.1 to 100 mg L^{-1} in the two extracts. GR gene was upregulated in oak (10 mg L^{-1} DOC) but not in reed extract. Elevated enzyme activities and overexpression of GR gene contributed towards combating the damaging effect of lipid peroxidation (LPO). Liquid chromatography and organic carbon detection (LC-OCD) analysis of leaf litter extracts showed the presence of a significant proportion of humic-like substances that were suggested to be one of the key components causing observed effects.

C. demersum displayed phenotypic plasticity in the two extracts. Acclimation to the effects of reed extract occurred after 7 days whereas in oak extracts, acclimation was not completed within the same period. In *L. minor*, acclimation to the two extracts was not evident within the period covered in the study. This suggests that biochemical and physiological acclimation requires different time-frames depending on the plant species and the stress factor.

The macrophytes manifested a notable level of antioxidative resilience depending on the plant species or the DOC source. Although biochemical defense mechanisms, to a large extent, averted oxidative damage (lipid peroxidation), inhibitory growth effects nevertheless occurred. It was concluded that oak and reed leaf litter has the potential to adversely affect aquatic macrophytes and probably structure aquatic ecosystems, especially at high DOC levels.

Zusammenfassung

Die Konzentration gelösten organischen Kohlenstoffs (DOC) in Gewässern wird von vielfältigen Faktoren beeinflusst, die von anthropogenen Störungen (z.B. durch die Landnutzung) bis hin zu wetterbedingten natürlichen Prozessen reichen. Zu diesen natürlichen Prozessen gehört die große Menge an Blättern, die von Laubbäumen jedes Jahr abgeworfen und zersetzt wird. Bei diesem Zersetzungsprozess gelangen chemische Substanzen in die Gewässer. Es wird geschätzt, dass das terrestrische Laubstreu bis zu 30% des gesamten DOC in Fließgewässern beiträgt. Während die Abbaurate des Laubstreus, deren Beeinflussung durch vielfältige ökologische Faktoren und der Einfluss auf den Kohlenstofffluss in aquatischen Systemen angemessen untersucht wurden, hat die potenzielle Auswirkung von Abbauprodukten auf aquatische Organismen bisher keine ausreichende Aufmerksamkeit auf sich gezogen.

In der vorliegenden Studie wurde Laubstreu-DOC von terrestrischen und aquatischen Pflanzen genutzt, um die Auswirkung der Abbauprodukte von Laubstreu auf die aquatischen Makrophyten *Ceratophyllum demersum* und *Lemna minor* zu untersuchen. Testparameter waren die Abgabe von photosynthetischem Sauerstoff, Enzyme des antioxidativen Systems, Glutathion-Redoxdynamik und die Wachstumsrate der Makrophyten. Die Zielsetzung war, das Potenzial von Laubstreu-DOC zu evaluieren, durch Veranlassung von oxidativem Stress, der eine Zellenzerstörung und Photosynthese- und Wachstumsreduzierung verursacht, auf aquatische Makrophyten einzuwirken. Dabei würde eine funktionelle Verknüpfung zwischen terrestrischen und aquatischen Ökosystemen aufgezeigt. Das Potenzial der untersuchten Makrophyten, sich an diese Auswirkungen während einer verlängerten Exposition (von 7 Tagen) anzupassen, wurde auch evaluiert.

Dabei wurden wasserhaltige Extrakte aus der Eiche (*Quercus robur*) und dem Schilf (*Phragmites australis*) als allochthone beziehungsweise autochthone Quellen von DOC genutzt. Dosis- und zeit-abhängige Expositionsexperimente wurden durchgeführt und verschiedene biochemische (enzymatische und nicht-enzymatische antioxidative Substanzen), physiologische (Photosynthese und Wachstum) und molekulare (Glutathion reduktase {GR} Genexpression) Parameter wurden gemessen.

Eine signifikante photosynthetische Hemmung wurde bei niedrigen und hohen DOC-Niveaus in Eichen- und Schilfextrakten beobachtet. Bei *C. demersum* konnte die photosynthetische Reduktion nicht durch Chlorophyllpigmente erklärt werden, weil diese nicht nachteilig beeinflusst wurden. Bei *L. minor* hingegen lässt sich die photosynthetische Hemmung mit der Reduzierung der Chlorophyllpigmente zumindest teilweise erklären, begleitet von einer Wachstumsminderung bei hohem DOC-Niveau.

Biotransformation und antioxidative Enzyme wurden bei DOC Konzentrationen von 0,1 bis 100 mg L^{-1} in beiden Extrakten induziert. Erhöhte GR Genexpression konnte bei den Eichenextrakten nachgewiesen werden (bei 10 mg L^{-1} DOC), jedoch nicht bei den Schilfextrakten. Erhöhte Enzymaktivitäten und übermäßige Expression der GR Gene trugen zur Vermeidung schädigender Effekte auf Lipid peroxidation (LPO) bei. Die LC-OCD (Liquid chromatography and organic carbon detection) Analyse der Laubstreu-Extrakte zeigte das Vorhandensein eines signifikanten Anteils humin-ähnlicher Substanzen, die als Schlüsselkomponenten für das Verursachen der beobachteten Effekte vorgeschlagen wurden.

C. demersum zeigte phenotypische Plastizität in beiden Extrakten. Bei den Schilfextrakten trat die Akklimatisierung nach 7 Tagen ein, wohingegen die Akklimatisierung bei den Eichenextrakten im gleichen Zeitraum noch nicht abgeschlossen war. Bei *L. minor* war während des gesamten Zeitraums der Studie keine Akklimatisierung erkennbar. Das deutet darauf hin, dass biochemische und physiologische Akklimatisierung verschiedene Zeitfenster voraussetzt, abhängig von der Pflanzenart und den Stressfaktoren.

Die Makrophyten zeigten ein beachtenswertes Niveau antioxidativer Widerstandsfähigkeit, abhängig von der Pflanzenart oder der DOC Quelle. Obwohl biochemische Abwehrmechanismen größtenteils oxidative Schädigungen (LPO) verhinderten, traten dennoch wachstumshemmende Effekte auf. Es wurde geschlussfolgert, dass Eichen- und Schilf-Laubstreu das Potential haben, aquatische Makrophyten negativ zu beeinflussen und mit hoher Wahrscheinlichkeit aquatische Ökosysteme, besonders bei hohem DOC Niveau, strukturieren.

Introduction

1.1 Background (Problem statement)

Many processes play an important role in the linkage between aquatic and terrestrial ecosystems. These processes include a complex system of biogeochemical cycling of materials and energy through food-web interactions, evapotranspiration, hydrochemical reactions and the transfer of carbon from terrestrial to aquatic systems (Grimm et. al., 2003). Natural organic matter (NOM) is a ubiquitous component of natural waters and affects the functioning of aquatic ecosystems. Aquatic ecosystems contain a wide range of NOM generated by the decay of, or exudation from, plant litter as well as associated microbial biofilms. Organic matter can be derived from terrestrial sources external to the aquatic system (allochthonous) and from sources within the aquatic system (autochthonous) (Meyer et al, 1998).

Figure 1. Terrestrial leaf-litter falls in and around aquatic systems, contributing to the total DOC pool (Photo by S. Pflugmacher).

In stream and river catchments, NOM is derived primarily from allochthonous sources such as organic soil horizons (Brooks et al, 1999), although autochthonous NOM forms a significant fraction of the total dissolved organic carbon (DOC) pool in aquatic systems (Cole et al,

2000). Jones (2005) indicated that nearly all freshwater systems contain some humic substances of allochthonous and certainly of autochthonous origin.
Leaves falling into water bodies constitute a significant proportion of organic matter input into the aquatic system (Fig. 1), contributing up to 30% of the total DOC pool in streams (Meyer et al, 1998). In many studies dealing with NOM, the DOC component is quantified. DOC concentrations in freshwater ecosystems vary widely, ranging between < 1 and up to 100 mg L^{-1} (Thurman, 1985; Wetzel, 2001), although higher concentrations (up to 300 mg L^{-1}) have also been reported in some Canadian wetlands (Blodau et al., 2004). DOC levels continue to increase in most waters due to various reasons ranging from anthropogenic impact (e.g. land-use systems) to natural processes related to weather phenomena (Evans et al. 2005). Contrary to the quantitative significance of DOC in freshwater ecosystems, there is still limited knowledge on their qualitative importance in terms of ecological functioning. The main focus has been on their direct or indirect role in the energy budget of aquatic bodies (eg. Sarvala et al., 1981). However, the biotic component and its interaction with the abiotic and other biotic factors in its surrounding constitute the most significant object of limnological study. For instance, the decay of leaf litter releases DOC, which interacts with aquatic organisms and produces reactive oxygen species triggering the antioxidative system, as otherwise the degradation of the system may occur.

The formation of reactive oxygen species and the consequences in the cell when faced with certain chemical substances is of great importance in the elucidation of essential questions in stress physiology. Modulation of antioxidative enzyme activity has been used as a convenient model for the investigation of stress originating from NOM, humic-like substances (Pflugmacher, 1999) as well as other environmental pollutants (Nimptsch et al., 2005). Leachates from decaying leaf litter could have adverse effects on aquatic organisms depending on the nature and chemical composition of the plant exudates at the different stages of decomposition (Nimptsch and Pflugmacher, 2008).

1.2 Sources of natural organic matter (NOM)

Natural organic matter constitutes a large number of organic compounds; hence it is practically impossible to provide a general description of their chemical nature. Nevertheless, NOM consists of low molecular weight compounds such as carbohydrates and amino acids, as well as complex high molecular weight compounds collectively described as humic substances (HS). HS are a complex mixture of aromatic and aliphatic hydrocarbon structures with amide, ketone, carboxyl and other functional groups attached (Leenheer and Croue, 2003). In freshwater bodies, most humic substances derive from terrestrial plant debris, of which lignins, tannins and terpenes (Fig. 2) are the main source material. These plant products are considered natural environmental chemicals in aquatic systems and make a connection between aquatic and terrestrial ecosystems.

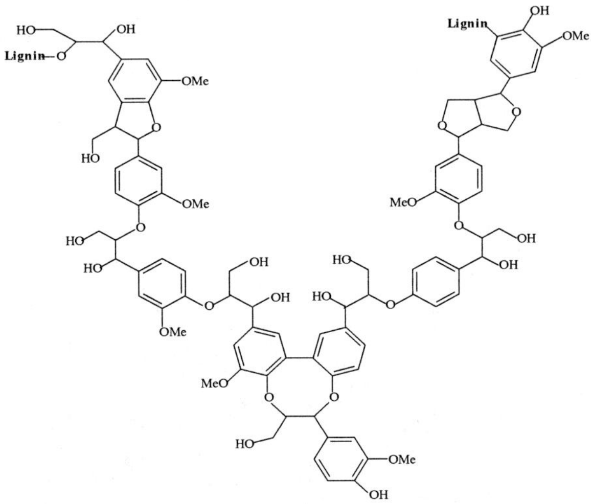

(i)

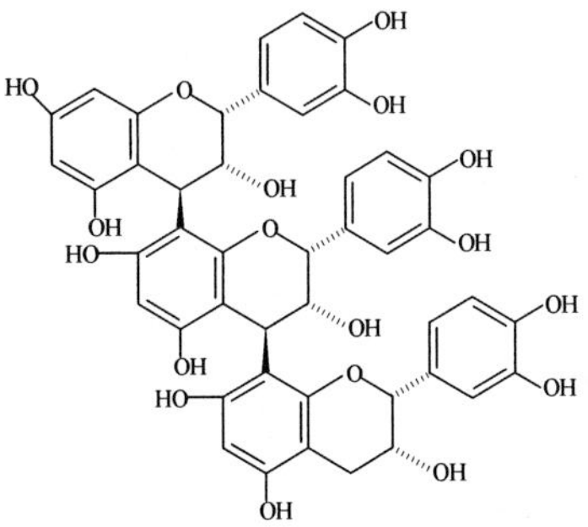

(ii)

(iii)

Figure 2. Chemical compounds with aromatic moities present in decaying plant debris (i) a piece of lignin polymer (ii) proanthocyanidin (tannin) (iii) α- Terpinene (terpene) (Source: http://www.ansci.cornell.edu/plants)

The main sources of natural organic matter in aquatic systems depend on the location and land-use patterns in adjacent terrestrial environment. Headwaters of lotic systems may be dominated by erosional and allochthonous inputs, whereas natural lake systems are mostly dominated by autochthonous inputs. Particulate organic matter in forested headwater streams and rivers is derived mainly from adjacent terrestrial vegetation and may contribute up to 800 g m^{-2} $year^{-1}$ (Ward et al., 1994). Leaf litter and woody debris constitute a significant component of the organic matter inputs and may reach the aquatic system directly by dropping from overhanging vegetation or may be transported from adjacent forest floor. In the present study, oak (*Quercus robur*) and reed (*Phragmites australis*) species were used as allochthonous and autochthonous sources of NOM respectively.

1.2.1 *Quercus robur* (oak) leaf litter

Q. robur leaf (Source: www.baumkunde.de)

Deciduous oak trees (e.g. *Quercus robur*), which form a significant proportion in deciduous and mixed forests in catchment areas, lose their leaves seasonally. Oak leaves therefore make an important allochthonous contribution to the organic matter input in streams, rivers and lakes. In a previous study, Salminen et al (2004) found hydrolysable tannins to be the dominant group of phenolic compounds in the leaves of the oak species *Quercus robur*. A rare dimeric ellagitannin, cocciferin D_2, was also detected for the first time in the leaves of *Q. robur*. A few earlier studies have shown that the leaves of *Q. robur* contain flavonol glycosides, castalagin, vescalagin and casuarictin (Scalbert et al, 1988; Grundhöfer et al, 2001).

1.2.2 *Phragmites australis* (reed) leaf litter

Phragmites australis is an emergent macrophyte, cosmopolitan in distribution and makes up one of the most productive natural plant population in the biosphere (Wetzel, 2001). It is the dominant and most conspicuous plant species in many aquatic systems in temperate regions and often forms mono-specific reed-stands and huge reed-beds along many shallow lakes (Hollis and Jones, 1991). It grows perennially, with a constant turnover of population members that are senescing as new cohorts are emerging and growing.

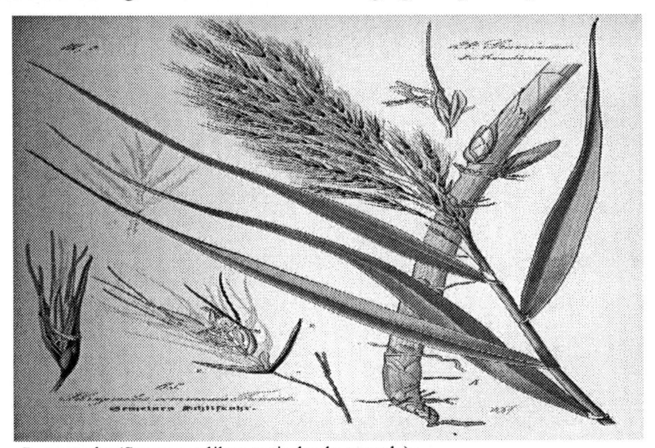

P. australis (Source: caliban.mpiz-koeln.mpg.de)

Cell membrane integrity collapses during senescence with major losses of organic matter into the aquatic system (Wetzel, 2001). *P. australis* thus constitutes a major autochthonous source of organic matter in shallow lakes and rivers. A recent reed decomposition study indicated that *P. australis* constitute an important autochthonous source of humic substances (HS) in water bodies rich in reed stand (V.-Balogh et al., 2006). NOM and HS from unknown sources have been implicated in oxidative stress induction and reactive oxygen species generation (Pflugmacher et al., 1999, 2003).

1.3 Reactive oxygen species (ROS) formation

Atmospheric oxygen in its ground-state has two unpaired electrons. This feature makes it usually non-reactive, unless activated, to organic molecules which have paired electrons with opposite spins. The production of reactive oxygen species (ROS) such as superoxide radicals, hydroxyl radicals and hydrogen peroxide may also result from metabolic reactions such as photorespiration, mitochondrial electron transport chain and from the photosynthetic apparatus. Furthermore, pathogens and wounding or environmental stresses (e.g. temperature, drought or osmotic stress) have been shown to trigger the active production of ROS (Dröge 2002, Penuelas et. al., 2005). In plant cells, superoxide anions are often formed by the reduction of molecular oxygen. This occurs either spontaneously by redox reactive compounds of the mitochondrial electron transport chain or by enzyme (e.g. xanthine oxidase) mediated processes. Other ROS are then generated from the superoxide anion radical in subsequent energy-independent reaction steps. The first step involves the protonation of the formed superoxide anion radical and the consequent formation of perhydroperoxyl radical.

$$H^+ + O_2^{\bullet -} \rightleftharpoons {}^{\bullet}HO_2$$

This may be followed by a spontaneous dismutation of a pair of perhydroperoxyl radicals to form hydrogen peroxide and molecular oxygen.

$${}^{\bullet}HO_2 + {}^{\bullet}HO_2 \longrightarrow H_2O_2 + O_2$$

Superoxide anion and perhydroperoxyl radical may also undergo spontaneous dismutation to form hydrogen peroxide, molecular oxygen and hydroxyl anion.

$${}^{\bullet}HO_2 + O_2^{\bullet -} + H_2O \longrightarrow H_2O_2 + O_2 + OH$$

The enzyme catalysed dismutation of two superoxide radicals by superoxide dismutases (SOD) also yield hydrogen peroxide.

$$2O_2^{\bullet -} + 2H \longrightarrow H_2O_2 + O_2$$

These partially reduced or activated derivatives of molecular oxygen are highly reactive and toxic and can therefore potentially lead to oxidative stress conditions or perform a signalling role in both plants and animals.

1.4 Oxidative stress and ROS signalling

Oxidative stress is a physiological phenomenon occurring due to circumstances in which critical cellular redox balance is disrupted. This often results from the overproduction and excess accumulation of ROS or depletion of antioxidants or both (Scandalios, 2005). In plants, ROS are produced endogenously during normal developmental transitions such as leaf and flower senescence, seed maturation as well as during normal photosynthetic and respiratory metabolism. When susceptible plants are exposed to xenobiotic or chemical substances, there is an overproduction of ROS, thereby causing oxidative stress conditions leading to damage of important biomolecules such as nucleic acids, proteins, lipids, pigments

(Fig. 3). The sugar and the base moieties of DNA are susceptible to oxidation, causing base degradation, single strand breakage, and cross-linking to protein (Imlay and Linn, 1986). A superoxide molecule is charged and thus cannot pass through biological membranes. Phase III of the biotransformation process involving subcellular compartmentalisation according to the so-called 'Green liver concept' is therefore crucial for the efficient removal and control of superoxide anions at their sites of production in the plant. H_2O_2 on the other hand is formed as a result of cellular biotransformation processes involving the action of superoxide dismutase (SOD), and is capable of diffusing across membranes and is thought to fulfill a signalling function in defense responses (Mullineaux et. al., 2000).

1.5 Mechanism of defense against oxidative stress in plants

Reactive oxgen species are formed in stressed as well as 'normal' habitat conditions. Under unstressed conditions, the production and removal of ROS are in dynamic equilibrium (Alscher et. al., 2002). However, when faced with increased generation of ROS during stress, the plants' defense system becomes overwhelmed and often results in the mobilisation of the antioxidant enzymes (e.g. superoxide dismutase, peroxidase, glutathione reductase, glutathione peroxidase, glutathione S- transferase) in order to prevent or minimise damage to cells (Fig. 3). These enzymes are predominantly located in the cytosol but also occur in various parts of the cell (Tab. 1).

Table 1. Subcellular locations of some antioxidant enzymes and molecules involved in plant defense

Antioxidant Enzyme/Molecule	Subcellular location
Superoxide dismutase	Cytosol, mitochodrion, plastid, peroxisome
Peroxidases	Cytosol, cell wall-bound
Catalase	Cytosol, mitochondrion, peroxisome, glyoxysome
Glutathione reductase	Cytosol, mitochondrion, plastid, chloroplast
Glutathione S-transferases	Cytosol, microsomal
Ascorbate peroxidase	Plastid stroma and membranes
Glutathione peroxidase	Cytosol, mitochondrion, chloroplast
Reduced glutathione (GSH)	Cytosol, mitochondrion, plastid
Tocopherol (vitamine E)	Cell and plastid membranes
Ascorbate (vitamine C)	Cytosol, plastid,apoplast, vacoule

Sources: Hausladen and Alscher, 1993; Mitova et al. (2004); Scandalios (2005)

Based on the so-called 'green liver' concept, Sandermann (1994) categorized plant metabolic sequences of biotransformation into three phases. Phase I involves transformation by hydrolysis and redox reactions and this is mainly achieved by cytochrome P-450 monoxygenases. This is then followed by conjugation reactions through the enzymes glutathione S-transferases and glucosyl transferases (phase II). Here the activated intermediate metabolites from phase I are then conjugated to the reduced tripeptide glutathione (GSH) or bound to sugar molecules. This leads to the formation of inactive water soluble conjugates. The third phase starts with the translocation and metabolisation of glutathione conjugates with peptidases and ends with internal compartmentalization in vacuoles and cell walls (phase III).

1.5.1 Enzymatic defenses

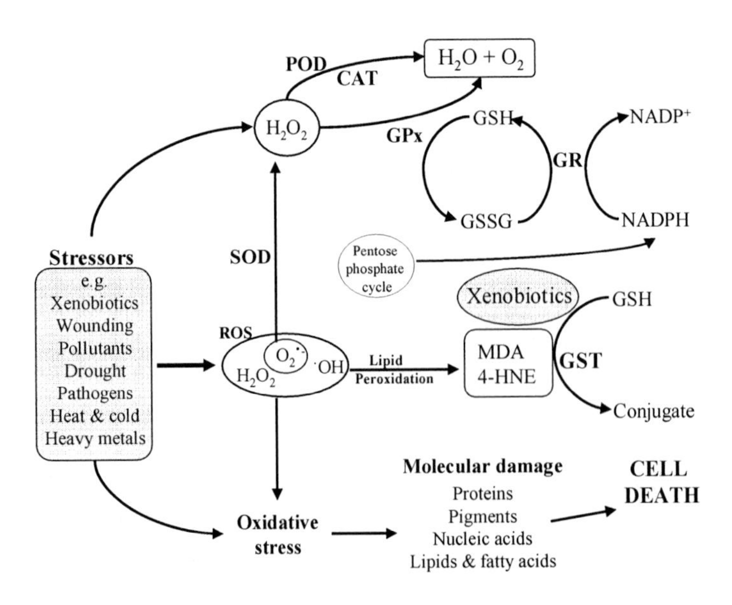

Figure 3. Schematic representation of stress-induced reactive oxygen species (ROS) production, biotransformation and antioxidative enzyme defense and toxicological consequences culminating to cell death in case of inadequate oxidative defense. Antioxidative enzymes - SOD: superoxide dismutase, CAT: catalase, POD: peroxidase, GPx: glutathione peroxidase, GR: glutathione reductase; Biotransformation enzyme - GST: glutathione S-transferase; Lipid peroxidation metabolites - MDA: melondialdehyde, 4-HNE: 4-hydroxenonenal; Antioxidants - GSH: reduced glutathione, GSSG: oxidized glutathione (Adapted from Scandalios, 2005)

1.5.1.1 Superoxide dismutases (SOD: EC 1.15.1.1)

The superoxide dismutases often form the first line of defense against excessive ROS formation within cells. Since superoxide radicals are potentially produced in different cell compartments even under normal metabolic functions, the SODs catalyse the dismutation of superoxide radicals to form hydrogen peroxide which is also reactive and toxic to cells. Alscher et al (2002) reported an up-regulation of different isoforms of SOD enzyme in *Arabidopsis thaliana* under xenobiotic stress.

1.5.1.2 Peroxidases (POD: EC 1.11.1.7)

The peroxidases are a group of enzymes, with many isoforms, involved in the next phase of the detoxification process. Their natural function is to prevent or minimize potential damage caused by hydrogen peroxide to plant cells and their constituents. Although hydrogen peroxide is formed during normal photosynthesis, the amount increases manifolds when plants are exposed to stress conditions, leading to the dismutation of the superoxide radical and formation of H_2O_2. Castillo et. al. (1987) used the activity of peroxidase enzymes as indicator of pollution stress in the leaves of some forest trees and in the roots of *Pinus sylvestris*. POD enzyme catalyses the reduction of hydrogen peroxide to water for detoxification. Some peroxidases can also reduce alkyl hydroperoxides to the corresponding alcohol (Fridovich, 1998).

1.5.1.3 Catalases (CAT: EC 1.11.1.6)

Catalases are tetrameric heme-containing enzymes with three known isoforms (Willekens et. al., 1994). Like the peroxidases, catalases enhance the conversion of hydrogen peroxide to water thereby protecting the cell from the damaging effect of H_2O_2 accumulation. However, the catalases are comparatively more efficient in dealing with relatively high concentrations of H_2O_2 because they exhibit a high K_m (in the millimolar range) for H_2O_2 (Fridovich, 1998; Scandalios, 2005). K_m being the concentration of substrate that leads to half-maximal enzyme velocity. Despite that they are efficient H_2O_2-scavengers, catalase enzymes can be photoinactivated in moderate light conditions to which plants are often adapted (Streb and Feierabend, 1996).

1.5.1.4 Glutathione reductase (GR: EC 1.8.1.7) and glutathione peroxidase (GPx: EC 1.11.1.9)

In plants, the tripeptide thiol, reduced glutathione (GSH) has facile electron donating capacity based on its sulphurhydryl group and thus acts as redox sensor to environmental perturbation. When plants are confronted with stress scenarios, the existing pool of GSH is converted to glutathione disulphide (GSSG) by the catalytic activity of glutathione peroxidase (GPx). GSH depletion may be the ultimate factor modulating vulnerability to oxidant attack. This can lead to the stimulation of glutathione biosynthesis on the one hand (Madamanchi et. al, 1994), or to the glutathione reductase (GR) catalysed and energy-dependent reduction of GSSG on the other. NADPH is the energy source for this reaction. GR occurs in various plant subcellular compartments such as chloroplasts, mitochondria and in the cytosol (Hausladen and Alscher, 1993).

1.5.1.5 Glutathione S-transferases (GST: EC 2.5.1.18)

Glutathione S-transferases (GSTs) are a large group of biotransformation enzymes involved in the conjugation of xenobiotics and other toxic substances exposed to plant and animal cells (Schröder, 1997; Menone and Pflugmacher, 2005). In plants, GSTs are found in the cytosol but may be also bound on membranes of cells and cell organelles. The membrane bound or insoluble GSTs are often referred to as microsomal GSTs. Different isoforms of cytosolic or soluble GST exist with varying capacities to detoxify xenobiotic compounds depending on the substrate specificity of the GST isoform and the species of organism involved (Nimptsch and Pflugmacher, 2005). The algicide 1-chloro-2,4-dinitrobenzene (CDNB) is one of the most commonly used model substrate for the non-specific testing of the biotransformation profile of GST enzymes in plants and animals.

1.5.2 Non-enzymatic antioxidant defenses

Compounds of intrinsic antioxidant properties including tocopherols, ascorbate and glutathione play an important role in maintaining intracellular redox balance in plant cells by the non-enzymatic scavenging of free radicals. Glutathione (GSH) is a tripeptide (γ-Glu-Cys-Gly) multifunctional metabolite which plays a key role in cellular defense and protection. It is the predominant form of free and reduced thiol in plants (Herschbach and Rennenberg, 1994) with the homologue, homoglutathione (hGSH), found mainly in legumes. GSH is known to prevent protein denaturation caused by oxidation of protein thiol groups during stress (Noctor et al., 2002). Additionally, GSH is an essential cofactor for antioxidant enzymes such GSTs and GPx. This is enhanced by the biochemically active nucleophilic sulphurhydryl group (-SH) (Fig. 4) of the cysteine moiety that reacts by nucleophilic displacement, thereby facilitating the excretion of xenobiotics from the organism's cell. Oxidation of the sulphurhydryl group often leads to the formation of the oxidised form of glutathione, glutathione disulphide (GSSG).

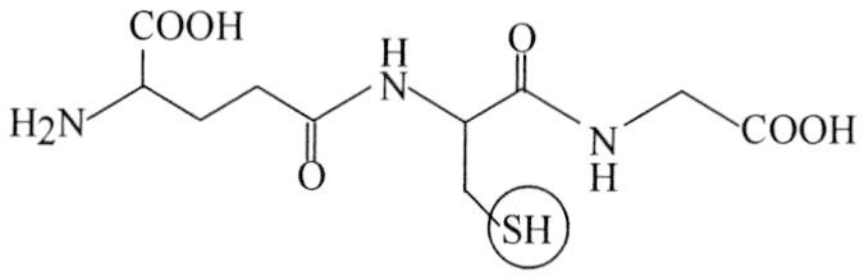

Figure 4. Reduced glutathione (GSH) showing the reactive sulphurhydryl group involved in GST-catalysed conjugation of xenobiotics and free radical scavenging mechanisms.

The balance between GSH and GSSG is homeostatically controlled by GSH biosynthesis and by the enzymes glutathione reductase (GR) and glutathione peroxidase (GPx). Therefore, GSH/GSSG ratio may be a sensitive indicator of oxidative stress. It is also suggested that the GSH/GSSG ratio may be more crucial in controlling gene expression and protein function than the absolute size of the glutathione pool (Noctor et al., 2002).

1.6 Lipid peroxidation

One of the consequences of overproduction of reactive oxygen species beyond the plant's defense capability is damage to cell membranes – described as lipid peroxidation. Polyunsaturated fatty acids of membrane phospholipids are highly susceptible to peroxidation by ROS. An autocatalytic chain of free radical reactions can produce various aldehydes, alkenals and hydroxyalkenals such as melondialdehyde (MDA) and 4-hydroxy-2-nonenal (4HNE) (Schneider et. al., 2001). The resulting alkyl radicals may rearrange to a more stable conjugated diene, which enters the autocatalytic lipid peroxidation cascade (Yang et. al., 2003). Membrane damage is mitigated by terminating the reaction chain through the removal of lipid hydroperoxides. Glutathione S-tranferases and glutathione peroxidase are the main enzymes involved in protecting the cells against ROS-induced lipid peroxidation.

1.7 Gene expression

Protein synthesis and gene expression involves the production and processing of messenger RNA (mRNA) in a multi-step process. This begins with the transcription of the nucleotide sequence coding for a particular protein into complementary mRNA molecules, followed by translation of the mRNA into proteins. Ultimately, folding and post-translational modifications of the gene product may occur. Various steps, including transcriptional and post-translational steps (e.g. phosphorylation), can be modulated. Genes whose expression is altered during periods of stress or adverse environmental conditions are presumed to be critical to an organism's survival (Scandalios, 2005). Evaluation of 'stress-responsive' genes is thus important for environmental quality assessment.

1.8 Photosynthesis and oxidative stress

Photosynthesis is an important anabolic process that takes place in the chloroplasts (thylakoid membrane) of plants as well as other autotrophic organisms such as algae and cyanobacteria. It consists of successive redox reactions during which light energy is absorbed by pigment complexes and transferred to the reaction centres of the photosystems (Fig. 5). Photosynthetic efficiency is an indicator of a plants' vitality or fitness in response to changes in their surrounding. The dynamics of the redox state of the chloroplasts as well as its electron transport system are important factors in the reactive oxygen species (ROS) signalling and ROS metabolic pathways in macrophytes (Baier and Dietz, 2005; Pflugmacher, 2006). Series of chain reactions in other cellular organelles such as the peroxisomes and the mitochondria also play a significant role in the overall metabolic homeostasis of ROS (Schiebe et al, 2005; Foyer and Noctor, 2005). It is thus obvious that photosynthesis, as a normal and essential metabolic process, also generates highly reactive oxygen species such as superoxide radicals which are by-products of reduction of molecular oxygen (Fig. 5). However, when plants are presented with xenobiotic compounds or influenced by other environmental factors, there is an enhanced production of ROS. This could lead to a stress condition that may influence the plant's photosynthetic efficiency (Bechtold et. al., 2005). Effects of environmental pollutants and toxicants on the photosynthetic activity of various algal species have been investigated (Kobbia et. al., 2001). Humic and/or humic-like substances could cause a change of the redox potential of components of the electron transport chain or of photosynthesis-coupled redox-active compounds (thioredoxin or glutathione) by affecting the electron transport efficiency (Pfannschmidt, 2003). Inhibition of photosynthesis by sorgoleone and sorgoleone-like compounds by disrupting the electron transfer chain between plastoquinone A (Q_A) and Q_B has been discussed (Weir et al, 2004). A similar mechanism is suggested for the effects of NOM on photosynthesis due to the presence of quinones and aromatic compounds (Fig. 5).

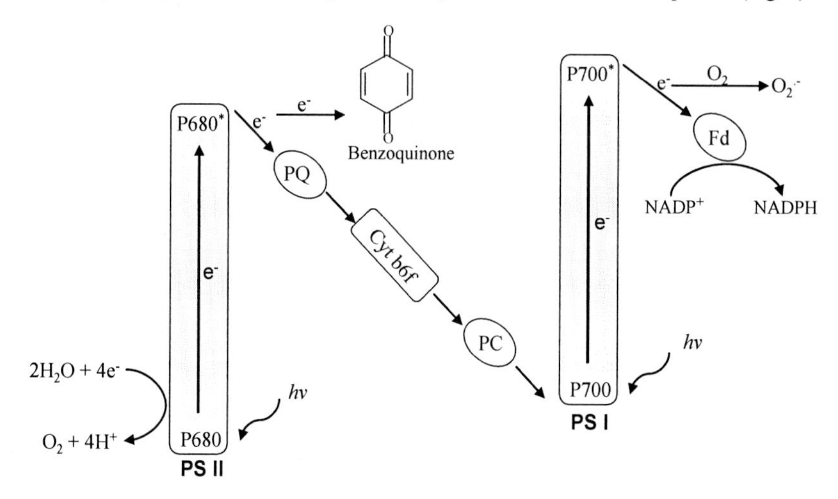

Figure 5. Z-scheme electron transport chain between photosystem II (PS II) and photosystem I (PS I). Chlorophyll molecule in PS II becomes excited by radiation (hv), then electrons from photolysis of water are transferred to the cytochrome b6f (Cyt b6f) complex by the electron carrier molecule plastoquinone (PQ). Electrons from the cytochrome b_6f complex are transferred to PS I by plastocyanin (PC). Aromatic structures (e.g. quinones) in leaf litter extracts are hypothesised to interfere with the electron transfer from PS II to PS I, thereby slowing down H_2O photolysis and hence oxygen release. Excited chlorophyll molecule at PS I transmits electrons through ferredoxin (FD) chain and eventually reduces $NADP^+$ to NADPH. Electrons of PS I can also be transfrerred to oxygen, which results in the generation of a superoxide radical, a reactive oxygen species. (Modified from: http://portalservices.org/biology/images)

1.9 Aquatic macrophytes

The distribution and abundance of aquatic plants are influenced by variations in environmental factors and hence serve as gauges of ecological integrity. Their sensitivity to short- and long-term changes in their surrounding makes them suitable indicators of environmental perturbations. There is growing concern about the dynamics of macrophytes and algae in aquatic systems. In most cases, studies have indicated that there has been a reduction in macrophyte communities and sometimes replaced by algal blooms (Körner 2001, 2002).

1.10 Acclimation

Plants and other aerobic living organisms have evolved various enzymatic and non-enzymatic antioxidant mechanisms for the regulation and control of ROS production and removal. These strategies enable plants to either reduce or mitigate the biological consequences associated with high ROS accumulation. However, when plants are subjected to extreme environmental conditions or exposed to pollutants or xenobiotics, there is usually an induced production of ROS above ambient levels, potentially causing oxidative stress. Sublethal exposure to mild stresses could initiate antioxidative defense processes that may confer some level of tolerance to further stress. This mechanism is suggested to play a key role in acclimation of plants to sudden exposure to adverse environmental conditions. Changes in the antioxidant enzymes, catalase and peroxidase, was associated with enhanced chilling tolerance of acclimated maize seedlings (Anderson et. al., 1995). In another study, Soleto and Khanna-Choppra (2006) clearly showed that drought acclimation was associated with induced changes in antioxidant enzymes and water relations in wheat seedlings.

1.11 Aim of study

The decomposition rate of leaf litter as influenced by various ecological factors and the contribution to carbon flux in the aquatic system has been aptly investigated (Gessner, 2000; Robinson and Gessner, 2000; Tiegs et al., 2008) but the potential implication for aquatic biota of the decomposition products has not received parallel attention.

The overall goal of this study was to assess the connectivity between aquatic and terrestrial ecosystems in terms of leaf litter-derived DOC release into aquatic systems and the accompanying physiological and biochemical consequences to aquatic plants. Therefore, the development of oxidative stress and photosynthesis in aquatic macrophytes when exposed to leaf-litter decomposition products was examined. Protective or acclimatory mechanisms employed to overcome oxidative stress were also investigated. For this purpose, leaf litter from the oak species, *Quercus robur*, was used as an allochthonous source of dissolved organic carbon (DOC). On the other hand, autochthonous litter input could also significantly contribute to the overall carbon flux in the aquatic system and influencing aquatic biota, and this was tested by using the reed, *Phragmites australis*, as a source of DOC.

Specifically, the study aimed at determining:

- whether and to what extent NOM from *P. australis* (Cav.) Trin Ex. Steud and *Q. robur* leaf litter affect photosynthesis and pigment contents in *C. demersum* and *Lemna minor.*
- the antioxidative defense status of these macrophytes after exposure to NOM.
- regulation of glutathione reductase (GR) gene expression after exposure to NOM
- acclimation potential of the macrophytes to the effects of NOM
- the growth response of *L. minor* after exposure to NOM

2. Materials and methods

2.1 Materials

2.1.1 Equipments

Table 2. List of equipments, products and software used in this study.

Equipment	Manufacturer
Photometer Uvikon^{TM} XL	Flowspek, 4057 Basel CH
Precision balance Research^{TM} R200D	Satorious, 37075 Goettingen
Ultra-Centrifuge Optima ^{TM} L-60	Beckman Coulter GmbH, 47807 Krefeld
Fixed angle-Rotor Type 70 Ti	Beckman Coulter GmbH, 47807 Krefeld
Plant vital 5000/1	INNO-Concept GmbH, 15344 Strausberg
Carbon analyzer TOC 5000A	Shimadzu Corporation
Gel Doc 2000	BioRad
Heat block	Neolab
PCR Master Cycler	Eppendorf
Sephadex G25 (DNA-grade, NAP-5)	Amersham Pharmacia
Microplate Flourometer, Spectrafluor Plus	Tecan, Crailsheim, Germany
Ultrasonic Processor UP 200s	dr. hielscher GmbH, Germany
Climatic Growth Cabintes	RUMED, Rubarth Apparate GmbH

2.1.2 Chemicals

Table 3. List of chemicals used in this study.

Chemicals	Company
Advanced solution	Cytoskeleton
Agarose	Cambrex
Ammonium sulphate ($(NH_4)_2SO_4$)	Sigma-Aldrich
Bovine serum albumin	Sigma-Aldrich
Bradford reagent	Sigma-Aldrich

Table 3 contd.

Bromophenol blue	Merck
Calcium chloride ($CaCl_2$)	Sigma-Aldrich
1-chloro-2,4-dinitrobenzene (CDNB)	Kodak
Chloroform	Roth
DEPC-water	Roth
Dimethyl sulphoxide (DMSO)	Sigma-Adrich
Disodium hydrogen phosphate (Na_2HPO_4)	Sigma-Aldrich
Dithioerythritol (DTE)	Sigma-Aldrich
5,5-Dithio-2-nitrobenzoic acid (DTNB)	Fluka
Ethanol (Absolute) (C_2H_5OH)	Merck
Ethidium bromide (EtBr)	Roth
Ethylene diamine tetraacetic acid (EDTA)	Sigma-Aldrich
Glutathione reductase (GR)	Sigma-Aldrich
Glycerol	Sigma-Aldrich
Glutathione (GSH)	Sigma-Aldrich
Glutathione disulphide (GSSG)	Sigma-Aldrich
Guaiacol	Sigma-Aldrich
Hydrogen peroxide (H_2O_2)	Aldrich
4-Hydroxynonenal (4HNE)	Alexis Biochemicals
Liquid nitrogen	Messner Griesheim
Magnesium chloride ($MgCl_2$)	Merck
Malondialdehyde (MDA)	Calbiochem
Molecular marker (100 bp DNA ladder)	Fermentas
Nicotinamide adenine dinucleotide phosphate reduced (NADPH)	Sigma-Aldrich
Potassium hydrogen carbonate ($KHCO_3$)	Sigma-Aldrich
Sea salt	Sigma-Aldrich
Sodium dihydrogen phosphate (NaH_2PO_4)	Sigma-Aldrich
Sodium hydrogen carbonate (NaHCO3)	Roth
Sucrose	Roth
Sulphosalicylic acid (SSA)	Sigma-Aldrich
Sulphuric acid (H_2SO_4)	Merck
Titanium (IV) chloride ($TiCl_4$)	Acros Organics

2.1.3 Plant materials

Ceratophyllum demersum (Hornwort or Coontail)

Source: http://newportal.gbif.org

Taxonomic hierarchy

Kingdom	Plantae
Phylum	Magnoliophyta
Class	Magnoliopsida
Order	Nymphaeales
Family	Ceratophyllaceae
Genus	*Ceratophyllum*
Species	*C. demersum* L.

Lemna minor (Common duckweed)

Source: http://en.wikipedia.org

Taxonomic hierarchy

Kingdom	Plantae
Phylum	Magnoliophyta
Class	Liliopsida
Order	Alismatales
Family	Lemnaceae
Genus	*Lemna*
Species	*L. minor* L

2.2 Methods

2.2.1 Preparation and chemical characterisation of leaf extracts

Standing dead *Phragmites australis* plants were collected from the littoral zone of Lake Müggelsee in April 2005. The above-ground plant parts were cut at about 1 m above the water level and composed of the leaf blade and culms. Senesced and fallen oak leaves (*Quercus robur*) were collected in the same period from the catchment area of Lake Müggelsee. These were air-dried in the lab for two days in order to maintain an approximately uniform humidity level and ground using a Homogeniser Mill (model 1094 Tecator). 200 grammes each of *P. australis* and *Q. robur* ground material were soaked separately in 3 L of medium containing de-ionized water, $CaCl_2$ (0.2 g L^{-1}), $NaHCO_3$ (0.103 g L^{-1}) and sea-salt (0.1 g L^{-1}) in separate plastic containers and stirred for 24 hours at room temperature. The resulting mixture was centrifuged (L-60 Ultracentrifuge, Beckman LL-TB-003A) at 20,000 × g for 10 minutes at 4°C to remove suspended materials. The supernatant was filtered using 0.45 µm pore cellulose-nitrate membrane filters (Sartorius AG, Germany) overlayed by 0.8 µm pore membrane filters. The membrane filters were first flushed with 1 L distilled water before use. In *L. minor* bioassay experiments, leaf extracts were filtered using 0.2 µm pore membrane filters in order to have sterile extracts. Total dissolved organic carbon (TDOC) of both leaf extracts was determined by high temperature combustion (TOC-5000A, Shimadzu, Australia) after removal of inorganic carbonates by acidification (DIN EN 1484, 1998). From the total organic carbon concentrations, serial dilutions of 0.1, 1, 10 and 100 mg L^{-1} DOC were prepared from both leaf extracts and used for exposure experiments.

Dissolved organic carbon (DOC) fractions were characterized by Liquid Chromatography and Organic Carbon Detection (LC-OCD) according to Huber and Frimmel (1996) and as described in detail by Sachse et al. (2001). Briefly, the chromatographable DOC portion of the filtered samples passes through a size-exclusion column (SEC) packed with resin (Toyopearl HW 50S, volume of 250 x 20 mm). Phosphate buffer (0.029 mol L^{-1}, pH 6.5) was used as eluent at a flow rate of 1 ml min^{-1}. Detectors were used for both DOC and absorbance (254 nm). DOC was detected by infrared (IR) absorbance of CO_2 after UV oxidation of DOC at 185 nm in a cylindrical UV thin-film reactor (Graentzel-reactor). Fractions were identified by using fulvic and humic acid standards from the International Humic Substances Society (IHSS). In the present study, they are summarized into three fractions: high molecular weight substances (HMWS, e.g. polysaccharides), humic or humic-like substances (HS) and low molecular weight substances (LMWS). The ratio between the spectral absorption coefficient (SAC in m^{-1}, at 254 nm) and the organic carbon of the humic fraction (DOC_{HS} in mg L^{-1}) was calculated as the aromaticity (L mg^{-1} m^{-1}).

2.2.2 Exposure to leaf-litter extracts

2.2.2.1 Dose-response exposures

Ceratophyllum demersum was purchased from Aqua Global (Dr. Jander & Co. OHG, Seefeld, Germany). Plants were first rinsed with medium water to remove adhering debris and cultivated non-axenically in Provasoli's medium containing de-ionized water, $CaCl_2$ (0.2 g L^{-1}), $NaHCO_3$ (0.103 g L^{-1}) and sea salt (0.1 g L^{-1}) in an aquarium two weeks prior to exposure experiments for acclimatization (Pflugmacher et. al., 2003). Supplementary light was provided by daylight lamps with an irradiance of 12µE/m^{2}s, a photoperiod of 14:10 hours light:dark cycle and temperature was maintained at 23 ± 1°C. In order to avoid artefacts of stress resulting from cutting of plant pieces, whole acclimatised plants ranging from 3 – 4 g FW were used. These were exposed separately to *P. australis* and *Q. robur* leaf extracts in serial concentrations of 0.1, 1.0, 10.0 and 100 mg L^{-1} DOC and placed under constant light and temperature (aquarium) conditions for 24 h (Fig. 6). In control experiments, only medium water (i.e. no leaf extract) was added. Each exposure concentration including control was replicated five times.

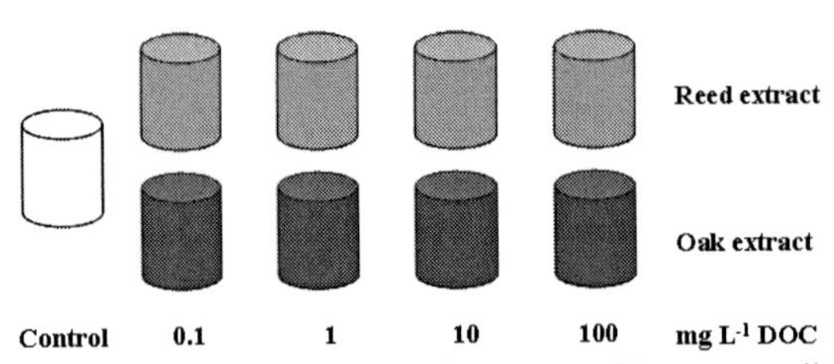

Figure 6. Dose response experimental layout for 24 h exposures. These were replicated 5 times in C. demersum. In *L. minor*, an additional DOC level (0.01 mg L^{-1}) was used inorder to obtain a no-effect-concentration, thus number of replicates was reduced to 4.

2.2.2.2 Time-response exposures

C. demersum was purchased from the same organisation and acclimatised under identical aquarium conditions as described for dose-response exposures. 3.5 ± 0.5g FW of acclimatised plants were exposed to 10 mg L^{-1} DOC of *P. australis* and *Q. robur* leaf extracts over durations of 1, 4, 8, 12, 24, 48 and 168 h respectively (Fig. 7). Zero (0) h control consisted of plants taken directly from the acclimatisation aquarium-tank, rinsed with medium water and immediately evaluated for photosynthetic oxygen production and pigments or shock-frozen for subsequent enzymatic assessments. Based on the assumption that plants may be influenced by differential time exposures to 'normal' experimental conditions, parallel controls (no leaf extract added to medium water) were respectively set up for each time exposure. Each exposure time was replicated four times.

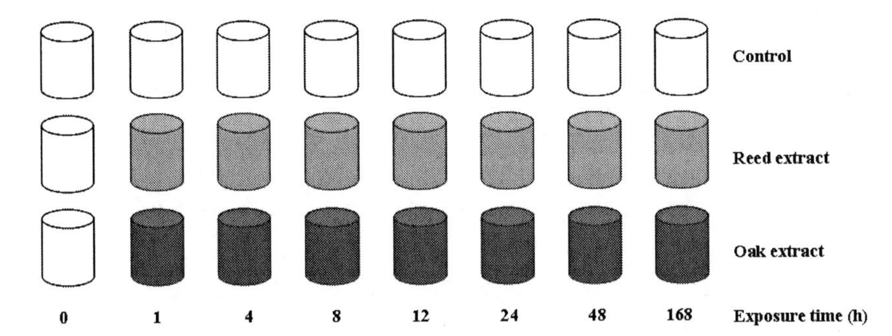

Figure 7. Time response experimental layout using 10 mg L^{-1} DOC of either oak or reed extracts. Zero (0) hour control represent samples taken directly from acclimatization tank and used for analyses without further exposure. In experiments lasting more than one day, leaf extracts were renewed every 24 h.

2.2.3 *Lemna minor* bioassay experiments

Experiments were performed on sterile monocultures of *L. minor* maintained within our laboratory (Leibniz Institute of Freshwater Ecology and Inland Fisheries, Biochemical Regulation) in modified Steinberg medium (Tab. 4) according to the guidelines of the organisation for economic cooperation and development (OECD 2006, draft 221). Prior to the test, subcultures were acclimated for 7 days to the experimental conditions. Tests were conducted in climatic exposure growth cabinets calibrated at 25 ± 1 °C, with white fluorescent tubes mounted on top. Light intensity was adjusted to 95 µE m^{-2} s^{-1} with a variability of less than 5 % in the test area. A non-reflecting surface was used in growth cabinet since test solution colour was found to influence growth of *L. minor* on a white surface (Cleuvers and Ratte, 2002). 12 fronds (4 colonies of 3 fronds) of similar size were aseptically placed in sterilised crystallising dishes containing 100 mL oak or reed extracts at nominal DOC concentrations of 0 (control), 0.01, 0.1, 1, 10 and 100 mg L^{-1} and covered with glass petri dishes. 10 times concentrated sterile Steinberg medium and sterile de-ionised water were used for preparing serial dilutions of leaf extracts. Handling procedures were done in laminar airflow cabinets (Heraeus, Hanau, Germany) under aseptic conditions. The experiment lasted for 7 days (168 h). To prevent nutrient limitation and depletion of DOC contents, a static-renewal method was employed whereby plants were aseptically transferred to freshly prepared serial leaf extract dilutions after 3 and 5 days of exposure, respectively.

The number of fronds in each replicate dish was recorded at the beginning of the exposure period (day 0) and on days 3, 5 and 7. Since plant biomass assessment involved destructive analysis, fresh and dry weight was determined on days 0 and 7 only. For this purpose, plants were harvested and the fresh weight immediately determined. Dry weight was determined by drying plants overnight at 60°C in an oven and weighed on a sensitive precision balance (Satorious, Goettingen, Germany). Initial (day 0) biomass was estimated on the day of exposure by determining the fresh and dry weights of 12 fronds (n = 8) from the same batch of *Lemna* plants used for the innoculum. Determination of chlorophyll pigment contents (see 2.2.5) also involved destructive analysis. Therefore, separate but identical exposures as above were conducted, which made possible the measurement of photosynthesis (see 2.2.4) and pigments throughout the experimental period i.e. at days 0, 3, 5 and 7.

Dose (24 h) and time (10 mg L^{-1} DOC) response exposures (see 2.2.2) were also conducted under identical conditions in growth cabinets for the assessment of biochemical parameters including glutathione, hydrogen peroxide contents and antioxidative enzyme activities (see 2.2.7 – 2.2.10).

Table 4. Chemical composition of stock solutions 1-5 used for the preparation of Steinberg medium as modified by Altenburger

Chemical substances	Concentration
1. Macroelements	g/L
Solution 1:	
KNO_3	17.5
KH_2PO_4	4.5
K_2HPO_4	0.63
Solution 2:	
$MgSO_4$ $7H_2O$	5.0
Solution 3:	
$Ca(NO_3)_2$ $4H_2O$	14.75
2. Microelements	mg/L
Solution 4:	
H_3BO_3	120
$ZnSO_4$ $7H_2O$	180
Na_2MoO_4 $2H_2O$	44
$MnCl_2$ $4H_2O$	180
Solution 5:	
$FeCl_3$ $6H_2O$	760
Triplex III (EDTA)	1500

Source: http://www.lemnatec.de/medien_faq.htm

Stock culture solution for the growth of *L. minor* was prepared by mixing 20 ml of stock solutions 1-3 and 1 ml of stock solutions 4 and 5 in a beaker and filled up to 1000 ml with nanopure water.

2.2.3.1 Determination of *Lemna* growth rate and inhibition

Frond number, fresh and dry weights were used as endpoints to determine average specific growth rates (μ) and to calculate growth rate inhibition according to OECD 221 guidelines.

$$\mu = \frac{(\ln x_{t2} - \ln x_{t1})}{t_2 - t_1}$$

where x_{t1} and x_{t2} represent the observation parameter at times t_1 and t_2 respectively.

Growth rate inhibition, based on frond number, fresh and dry weight, as a function of DOC concentration was described by a four-parameter sigmoidal dose-response model:

$$Y = \frac{\text{Bottom} + (\text{Top- Bottom})}{1+10^{(\text{LogEC50 - X})\cdot\text{Hill slope}}}$$

where X is the logarithm of concentration, Y is the response which starts at Bottom (0 %) and goes to Top (100 %) with a sigmoid shape (GraphPad Prism version 4.03 for windows, GraphPad software, San Diego, California, USA). Median growth effect (EC_{50}) for nominal DOC concentrations was calculated for the various endpoints after non-linear regression curve fitting.

2.2.4 Measurement of photosynthesis

Photosynthetic oxygen production of the macrophytes was measured using a Plant Vital 5000 (Inno-Concept, Straussberg) device. The device detects and records oxygen concentration within the vicinity of the specially devised electrochemical sensor at 100% light intensity (1 klux) and a constant temperature of 20°C. Measurements were made in a 0.5 M $KHCO_3$ solution. Prior to measuring, plants were rinsed with medium water in order to mitigate the effect due to the colour of the leaf extracts. To keep the delay time between removing the plant material from the exposure medium and measuring to the barest minimum, plants were weighed only after measuring photosynthesis. Maximum recorded rate of photosynthetic oxygen production was expressed as µmoles O_2 * s^{-1} * g^{-1} FW. After measuring photosynthesis and taking samples for pigment analysis, remaining plant materials were immediately frozen in liquid nitrogen and kept at -80 °C until enzyme preparation.

2.2.5 Pigment analysis

Chlorophyll pigments were extracted and analysed according to Inskeep and Bloom (1985). 25 mg (*C. demersum*) or 10 mg (*L. minor*) of plant material was completely immersed into 1.5 ml N,N-dimethylformamide (N,N-DMF) and kept in the dark for 48 h at room temperature to allow for pigment extraction. The pigment extract was then transferred into 1.5 ml microcentrifuge tubes and centrifuged for 3 min at 3,000 rpm. The absorbance of the supernatant was measured at wavelengths of 664.5 and 647 nm. Handling and measurement of pigment samples was carried out in semi-dark in order to avoid the influence of light on the pigments. Chlorophylls *a* (Chl *a*), *b* (Chl *b*) and total chlorophyll (*t*Chl) contents were calculated and expressed as milligrams per gram fresh weight based on the extinction coefficients determined by Inskeep and Bloom (1985) as follows:

$$\text{Chl } a = (12.7 \times ABS_{664.5}) - (2.79 \times ABS_{647})$$

$$\text{Chl } b = (20.7 \times ABS_{647}) - (4.62 \times ABS_{664.5})$$

$$t\text{Chl} = (17.9 \times ABS_{647}) + (8.08 \times ABS_{664.5})$$

2.2.6 Antioxidant (GSH and GSSG) determination

Extract preparation and assay for the determination of total (GSH + GSSG) and oxidised (GSSG) glutathione contents in the macrophytes was done according to the methods described by Anderson (1985). Ground plant tissue material was quickly homogenized in 5 ml of 5 % sulphosalicylic acid. The homogenate was centrifuged at 18,000 × g for 12 min at 4 °C to remove cell debris. 400 µl portion of the supernatant (crude extract) was used for total glutathione assay. Another 400 µl portion of the same crude acid-extract was flash frozen in liquid nitrogen and used for further GSSG extraction. 20 µl of 2-vinylpyridine was added and incubated for 1 h at room temperature (25°C). Vinylpyridine binds with and prevent auto-oxidation of reduced glutathione (GSH). The mixture was centrifuged at 14,000 × g for 10 min at 4°C and the supernatant was used for GSSG assay. The concentrations of total glutathione (GSH + GSSG) and GSSG contents were spectrophotometrically determined by an enzyme-recycling assay. The assay was based on sequential oxidation of glutathione by 5,5'-dithio-2-nitrobenzoic acid (DTNB) and reduction by NADPH in the presence of known amount of glutathione reductase (GR). The reaction was initiated by adding GR and the colour development of TNB was measured at 412 nm. Standard curves were made using pure GSH and GSSG and were used for calculating the amount of total glutathione (GSH plus GSSG) and GSSG present in samples. Reduced glutathione (GSH) was calculated as the difference between total glutathione and GSSG. Calculations were based on GSH equivalents according to the stoichiometric proportions of GSH and GSSG. The assay mixture for total glutathione in disposable cuvettes consisted of:

Reagent	Volume (µl)
Na_3PO_4-buffer (143 mM, pH 7.5; 6.3 mM EDTA)	840
5,5'-dithio-2-nitrobenzoic acid (DTNB) (6mM)	100
NADPH	25
Sample/Buffer (blank ref.)	10
Glutathione reductase (GR)	25

2.2.7 Intracellular hydrogen peroxide determination

Cell internal H_2O_2 content was determined according to Jana and Choudhuri (1982). 1 ± 0.3 g plant tissue was homogenized in 3 ml of 50 mM sodium phosphate buffer (pH 7.0). The homogenate was centrifuged at 19100 × g for 5 min at 4 °C. 250 µl of the supernatant was mixed with 750 µl of 0.1% titanium sulphate in 20% (v/v) H_2SO_4 and the absorbance measured after 1 min at 410 nm. H_2O_2 content was calculated using the extinction coefficient 0.28 l $mmol^{-1}$ cm^{-1} and expressed as nmol g^{-1} FW. The assay mixture, in disposable cuvettes, consisted of the following:

Reagent	Volume (µl)
0.1 % Titanium chloride in 20 % H_2SO_4	750
Sample/H_2O (blank ref.)	250

2.2.8 Multi-enzyme preparation

Enzyme preparation was done according to the method described by Pflugmacher and Steinberg (1997) (Fig. 8) with slight modification on the steps leading to the microsomal fraction. The frozen plants were ground to a fine powder in liquid nitrogen. The ground powder tissue was suspended in sodium phosphate buffer (0.1M, pH 6.5) containing 20% glycerol, 1.4 mM dithioerythritol (DTE) and 1 mM ethylenediamine tetraceticacid (EDTA) and homogenised using a glass potter. Cell debris was removed by centrifuging the slurry at 10,000 × g for 10 min. The residue was suspended in the same buffer and centrifuged again at 5000 × g for 5 min. The supernatant from the first and second centrifugation was then centrifuged at 40,000 × g for 60 min. The pellets, containing membrane fractions and defined as the microsomes, were suspended in 1 mL sodium phosphate buffer (20 mM, pH 7.0) containing 20% glycerol and homogenized using glass potter. The resulting 1mL microsomal protein fraction was immediately transferred into 1.5 mL eppendorf tubes and shock-frozen in liquid nitrogen before storage at -80 °C. The supernatant was precipitated with solid ammonium sulphate in two saturation steps, 0-35% and 35-80%, followed by centrifugation at 20,000 × g for 20 minutes and at 30,000 × g for 30 minutes respectively. The precipitate from the second precipitation which contains soluble protein-pellets was re-suspended in 0.5 mL sodium phosphate buffer (20 mM, pH 7.0) and desalted by gel filtration using NAP-5 columns (Amersham Pharmacia, Germany) which had been equilibrated with cytosolic buffer (Na_3PO_4 20 mM, pH 7). The eluted and now purified cytosolic (soluble) protein fraction (1 mL) was also collected in 1.5 mL eppendorf tubes and immediately shock-frozen in liquid nitrogen and stored at −80 °C for enzyme activity assay.

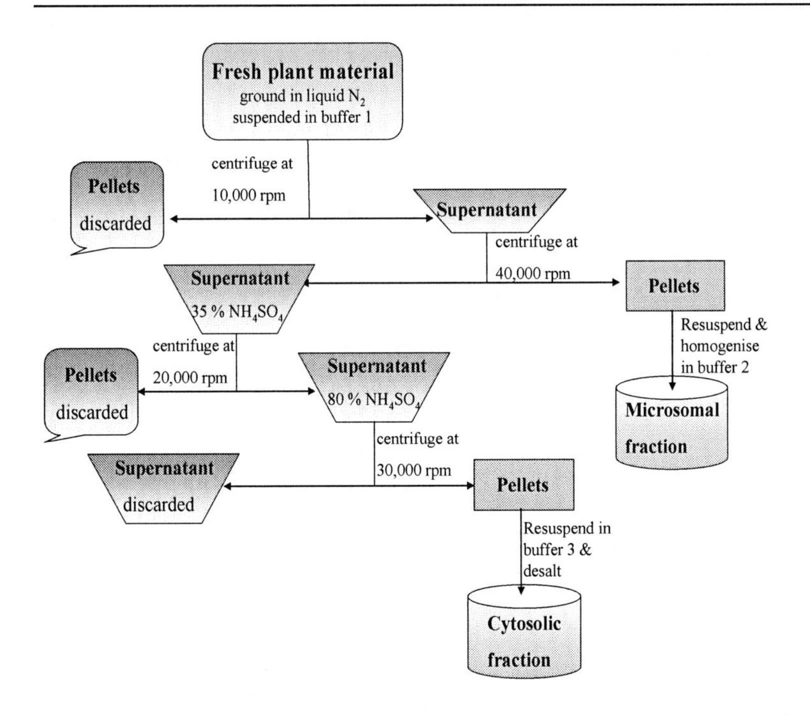

Figure 8. Scheme describing the stepwise procedures for the cytosolic and microsomal multienzyme preparation according to Pflugmacher and Steinberg (1997).

2.2.9 Determination of protein concentration

The protein content of each sample was determined by the method of Bradford (1976) using the Bradford protein dye reagent (Sigma). However, when protein contents in samples were below detection limits for Bradford reagent, a more sensitive 'Advanced Protein Assay (ADV01)' using 'Advance Solution' from Cytoskeleton was employed. The protein dye in this reagent solution is five times more concentrated than in Bradford reagent. The disadvantage of the Advance protein method is that a relatively large amount (800 µl compared to 25 µl per cuvette) of sample material was required but this is compensated by the very high sensitivity of the reagent making it possible to dilute samples up to a factor of 50 times. Nonetheless, both assay methods are based on the same principle of a change in maximum absorbance from a wavelength of 465 nm to 595 nm of Coomassie Brilliant Blue G-250 which is measured spectrophotometerically. Bovine serum albumin (BSA 98%, Sigma) was used as standard for calibration in both cases. A new calibration curve was prepared for every experiment. The protein assay mixture was composed of:

	Reagent	Volume (µl)
Bradford ($\lambda_{595\ nm}$)		
	Bradford reagent	1226
	Sample/H_2O (blank ref.)	25
Advanced ($\lambda_{590\ nm}$)		
	Adv01 reagent	100
	Sample/H_2O (blank ref.)	800

2.2.10 Biochemical assays

2.2.10.1 Superoxide dismutase activity assay

Superoxide dismutase (SOD: enzyme commission (EC) number 1.15.1.1) activity was determined by colorimertic measurement at 450 nm according to the user protocol of the SOD Assay Kit (Fluka, Buchs, Switzerland) based on xanthine/xanthine oxidase inhibition. The method utiizes a highly water-soluble tetrazolium salt, 2-4-iodophenyl-3-4-nitrophenyl-5-2,4-disulphophenyl-2H-tetrazolium monosodium salt. This salt produces a water-soluble formazan dye upon reduction with a superoxide anion. The rate of reduction is linearly related to the xanthine oxidase activity and is inhibited by SOD. One unit of SOD activity is defined as the amount of enzyme that caused 50 % inhibition of the control rate of tetrazolium salt reduction. For SOD, the assay consisted of:

Reagent	Volume (µl)
Dilution buffer	20
Tetrazolium salt	200
Sample/H_2O (blank ref.)	20
Xanthine oxidase enzyme	20

$$2O_2{\cdot}^- + 2H^+ \xrightarrow{SOD} 2H_2O_2 + O_2$$

(↓Abs = 450 nm)

2.2.10.2 Catalase activity assay

Catalase (CAT: EC 1.11.1.6) activity was assayed photometrically following the method of Baudhuin et al. (1964), based on the decrease of H_2O_2 corresponding to a decrease in absorbance measured at 240 nm (extinction coefficient $\varepsilon = 0.0361\ l\ mmol^{-1}\ cm^{-1}$). The composition of the assay mixture was as follows:

Reagent	Volume (µl)
Na_3PO_4-buffer (50 M, pH 7.0)	1250
H_2O_2 (150 mM)	100
Sample/H_2O (blank ref.)	25

$$2H_2O_2 \xrightarrow{CAT} 2H_2O + O_2$$

(↓Abs = 240 nm)

2.2.10.3 Peroxidase activity assay

Peroxidase (POD: EC 1.11.1.7) activity in the soluble fraction was measured spectrophotometrically using guaiacol as substrate (Drotar et. al., 1985; Bergmeyer, 1986). The principle of this assay is that guaiacol is oxidised in the presence of hydrogen peroxide and polymerizses to octahydrotetraguaiacol followed by a change in the colour and absorbance which is monitored at a wavelength of 436 nm (extinction coefficient $\varepsilon = 25.5\ l.\ cm^{-1}\ mM^{-1}$). Measurements were made in triplicates along with a blank reference over 5 minutes duration. Samples were diluted fifty times using cytosolic buffer (NaP 20 mM, pH 7.0). The assay mixture was made of the following:

Reagent	Volume (µl)
Na_3PO_4-buffer (0.1 M, pH 5.0)	860
Guaiacol solution (DMSO)	40
H_2O_2 (200 mM)	40
Sample/H_2O (blank ref.)	40

$$\text{Guaiacol} + 2H_2O_2 \xrightarrow[\text{pH } 5.0]{\text{POD}} \text{Tetraguaiacol} + 2H_2O + O_2$$

OH, OCH3

Guaiacol

(↓Abs = 240 nm)

2.2.10.4 Gutathione reductase (GR) activity assay

Glutathione reductase (GR, EC 1.6.4.2) activity in the soluble fraction was measured spectrophotometrically according to Carlberg and Mannervik (1985) based on the oxidation of NADPH to $NADP^+$ which is accompanied by a decrease in absorbance at a wavelength of 340 nm (extinction coefficient $\varepsilon = 6.4\ l\ mmol^{-1}\ cm^{-1}$). In the process, glutathione disulphide (GSSG) present in the reaction mixture is reduced by GR to GSH. The reaction mixture consisted of:

Reagent	**Volume (µl)**
Na_3PO_4-buffer (0.1M, pH 7.5)	850
Glutathione disulphide (GSSG) (20 mM)	50
NADPH (2 mM)	50
Sample/H_2O (blank ref.)	50

$$\text{GSSG} + \text{NADPH} + H^+ \xrightarrow[\text{pH } 7.5]{\text{GR}} 2\ \text{GSH} + NADP^+$$

GSSG

GSH (↓Abs = 340 nm)

2.2.10.5 Glutathione peroxidase (GPx) activity assay

GPx activity was followed spectrophotometrically over 3 minutes duration at 340 nm as described in Drotar et al. (1985). GPx samples were added to the test mixture at room temperature, and the NADPH oxidation rate was monitored. The reaction was initiated by the addition of hydrogen peroxide (9 mM) which is reduced by GPx as GSH becomes oxidized to GSSG. Since GR is also present in reaction mixture, GSSG is again reduced to GSH while NADPH becomes oxidised to yield $NADP^+$ thereby causing a decrease in absorbance at a wavelength of 340 nm (extinction coefficient $\varepsilon = 6.4\ l\ mmol^{-1}\ cm^{-1}$). Activity was calculated from the rate of NADPH oxidation. The assay composition consisted of:

Reagent	Volume (µl)
Na_3PO_4-buffer (0.1 M, pH 7.5)	800
Glutathione reductase (GR) (2U)	10
GSH (50 mM)	40
NADPH (2 mM)	40
H_2O_2 (9 mM)	10
Sample/H_2O (blank ref.)	100

$$2\ \text{GSH} + 2NADPH + 2H_2O_2 \xrightarrow[\text{pH 7.5}]{\text{GPx}} \text{GSSG} + 2NADP + 4H_2O$$

GSH GSSG (↓Abs = 340 nm)

2.2.10.6 Glutathione S-transferases (GST) activity assay

Activity of glutathione S-tranferases (GST) in soluble (cytosol) and microsomal fractions was assayed with 1-chloro-2,4-dinitrobenzene (CDNB) as substrate according to Habig et. al. (1974). Furthermore, GST activity in the cytosol was measured with 4-hydroxynonenal (4-HNE) as substrate according to (Alin et al., 1984). The reaction is initiated in each case by addition of the enzyme to the assay mixture.The principle is based on the enhanced conjugation of the algicide (CDNB) or the lipid peroxidation product (4-HNE) on the sulphurhydryl (-SH) group of the co-substrate glutathione (GSH) present in the reaction mixture. The change in absorbance resulting from the formed conjugates was measured at 340 nm (extinction coefficient $\varepsilon = 9.6\ l\ mmol^{-1}\ cm^{-1}$) and 224 nm (extinction coefficient $\varepsilon = 14{,}400\ l\ mmol^{-1}\ cm^{-1}$) for CDNB and 4-HNE respectively. The GST assay mixture respectively consisted of the following:

	Reagent	Volume (µl)
GST measured with CDNB		
	Na_3PO_4-buffer (0.1 M, pH 6.5)	1100
	Glutathione (GSH) (58.6 mM)	40
	CDNB (29.6 mM)	40
	Sample (enzyme)/H_2O (blank ref.)	40
GST measured with 4-HNE		
	Na_3PO_4-buffer (0.1 M, pH 6.5)	850
	Glutathione (GSH) (10 mM)	50
	4-HNE (2mM)	50
	Sample (enzyme)/H_2O (blank ref.)	50

Cl, NO_2, NO_2 + GSH —GST, pH 6.5→ HO, NH_2, O, O, H, N, O, H, H, N, H, O, OH, S, S, NO_2, NO_2 + HCl

1-chloro-2,4-dinitrobenzene (CDNB) GS-DNB conjugate (↓Abs = 340 nm)

O, OH + GSH —GST, pH 6.5→ O, SG, OH

4-Hydroxynonenal (4-HNE) GS-4NHE conjugate (↓Abs = 224 nm)

2.2.10.7 Calculation of specific enzyme activity

Specific activity is a convenient method of communicating the amount of enzyme activity present in each milligram of protein. The catalytic activity of an enzyme is determined by the rate (velocity) of the catalysed reaction under optimal reaction conditions (optimal pH, temperature and substrate concentration high enough that maximum rate is reached). The rate of the reaction is defined by the substrate turn over per unit time and expressed as micromoles per second (µmol/s). The catalytic activity (CA) in a specific volume, which is a specific factor, was thus calculated as follows:

$$CA = \frac{\Delta e \times V \times 1000}{\varepsilon \times d \times \Delta t \times v}$$

Δe = extinction change per minute; V = final volume of enzyme assay (µl); ε = substrate molar extinction coefficient ($1\ mmol^{-1}.cm^{-1}$) ; d = cuvette width (cm); Δt = measuring time interval; v = volume of enzyme extract in assay (µl)

For the purpose of comparing enzyme activity among treatments, the catalytic enzyme activity (CA) was expressed as a factor of the total protein concentration in the enzyme extract thereby yielding the specific enzyme activity (SA).

$$SA = \frac{CA}{[Protein]}$$

[protein] = protein concentration of enzyme extract ($\mu g\ \mu l^{-1}$)

A conventional unit for expressing enzyme activity is called the katal. One katal is the amount of enzyme that catalyses the conversion of one mole of substrate per second. Enzyme activity was thus uniformly expressed in nanokatals per milligram protein (nkat mg^{-1} protein).

The multi-enzyme extracts were constantly kept on ice during all assays and were stored at -80°C after shock-freezing them in liquid nitrogen.

2.2.11 Measurement of lipid peroxidation

Lipid peroxidation (LPO) was determined according to the User protocol of the lipid peroxidation assay kit (CALBIOCHEM, Cat. No. 437634) with minor modifications based on the method of Botsglou et al. (1994). The aldehyde, malondialdehyde (MDA), is an end product derived from the breakdown of lipid peroxides, which are often unstable. Estimation of MDA concentration serves as an index of lipid peroxidation. The reaction between N-methyl-2-phenylindole and MDA produces a stable chromophore with maximum absorbance at 586 nm. Plant material was weighed, homogenized in sodium phosphate buffer (20 mM, pH 7.0) and centrifuged for 10 minutes at 14000 x g and at 4°C. In order to prevent sample oxidation, 10 µL of 0.5 mM butylated hydroxytoluene (BHT) was added. The supernatant was then used for the assay. 650 µL of N-methyl-2-phenylindole in acetonitrile was added to 200 µL of the sample and gently mixed. 150 µL of 2N HCl was then added, mixed again and incubated for 1 h at 45°C. Samples were cooled on ice and the absorbance measured at 586nm. An MDA standard curve was established and used to determine the MDA content in samples.

Malondialdehyde + 2 N-methyl-2-phenylindole → Chromophore (Abs_{max} = 586 nm)

2.2.11.1 Calculation of Lipid peroxidation

Net absorbance at 586 nm (A_{586net}) was calculated for each standard using the standard curve data as:

$$A_{586net} = A_{586} - A_0$$

A_{586} = Absorbance at 586 nm
A_0 = Absorbance of blank sample

A586net was plotted against known MDA concentration from standard curve data and a linear regression analysis of A_{586net} on [MDA] was performed:

$$A_{586net} = a\,[MDA] + b$$

Therefore, unknown MDA concentration in sample was calculated as

$$[MDA] = \frac{(A_{586net} - b) \times df}{a}$$

[MDA] is the concentration of MDA in the sample

A_{586net} = net absorbance at 586 nm of the sample

a = regression coefficient (slope)

b = intercept

df = dilution factor

2.2.12 Molecular biology experiment (GR gene isolation and expression)

2.2.12.1 RNA isolation

Total RNA isolation from *C. demersum* was done according to the user protocol of the RNeasy Plant Mini Kit (Qiagen, Hilden, Germany) with slight modification on the steps leading to the disruption of the plant cell wall. Exposed plants were quickly rinsed with fresh medium water to remove debris and leaf extracts adhering on the surface and immediately flash-frozen in liquid nitrogen. Tips of plants were grounded to a fine powder in liquid N_2 using sterilised and cooled mortar and pestle. 100 mg powdered plant material was transferred to sterile15 ml tubes and 450 µl lysis buffer RLC containing 14.3 M β-mercaptoethanol was added and vigorously vortexed. In order to enhance greater cell wall disruption and to maximise RNA yield, ultrasonification (Ultrasonic Processor UP 200s, dr. hielscher GmbH) was carried out at amplitude 75 (cycle 0.5) for 30 secs. The resulting lysate was transferred to a QIA shredder spin column and centrifuged at 13,400 rpm for 2 mins. The supernatant of the flow-through was transferred into a 1.5 ml sterile eppendorf tube and 70% ethanol (1:2 v/v, ethanol:supernatant) was added and gently mixed. The mixture was transferred to RNeasy mini columns placed on 2 ml collection tubes and centrifuged at 13,400 rpm for 30 secs. The flow-through was discarded and 350 µl of buffer RW1 was added and centrifuged at 13,400 rpm for 30 secs. 80 µl of DNAse mixture (10 µl DNAse I solution + 70 µl RDD buffer) was applied on the membrane of the columns and incubated for 15 mins to ensure the digestion and removal of residual DNA. The columns were washed with 500 µl RPE buffer and centrifuged at 13,400 rpm for 30 secs. Washing of columns was repeated with 500 µl RPE buffer and centrifuged at 13,400 rpm for 2 mins. The columns were then eluted with 30 µl RNAse free water and the RNA solution was collected on sterile 1.5 ml eppendorf tubes. Total RNA content was quantified spectrophotometrically using a microplate fluorometer (Spectrafluor Plus, Tecan, Crailsheim, Germany).

2.2.12.2 Reverse transcription

Total RNA isolated was adjusted to 0.1 µg/µl using RNase-free DEPC (diethyl pyrocarbonate)-treated water. 1 µg DNase treated RNA was used for cDNA synthesis by incubating with 0.8 µl (0.5 µg/µl) Oligo $(dT)_{12\text{-}18}$ primers (Invitrogen) at 70°C for 5 mins. The reverse transcription was performed using 3 µl RT buffer (5×), 0.8 µl dNTPs (10mM) and 0.4 µl M-MLV reverse transcriptase (Promega) and incubated at 42 °C for 1h. The process was terminated by heating at 94 °C for 4 mins. Resulting cDNA samples were stored at -20 °C.

2.2.12.3 Semi-quantitative RT-PCR

cDNA specific for glutathione reductase (GR) and Glyceraldehyde-3-phosphate dehydrogenase (GAPDH) in *Ceratophyllum demersum* were amplified by semi-quantitative RT-PCR. PCR reaction for GAPDH expression consisted of 2 µl of cDNA sample, 14.7 µl DEPC-treated nanopure water, 2 µl PCR-buffer, 0.4 µl forward and reverse GAPDH primers, 0.4 µl dNTPs (10 mM) and 0.1 µl HotStar Taq DNA polymerase (Qiagen). Reaction mixture for GR amplification PCR consisted of 5 µl cDNA sample, 10.6 µl DEPC-treated nanopure water, 2 µl Pfu-buffer, 0.8 µl $MgCl_2$ (25 mM), 0.4 µl forward and reverse degenerate GR primers, 0.4 µl dNTPs (10 mM) and 0.4 µl Pfu DNA polymerase (Promega). Amplification was performed in a Thermal Mastercycler (Eppendorf, Hamburg, Germany) with the following thermal cycling conditions for GR: 2 mins at 95°C, 30 s at 95°C, 30 s at annealing temperature (55°C), 1 min at 72°C and finally 10 mins at 72°C. The PCR settings for GAPDH were 15 mins at 94°C, 50 s at 94°C, 50 s at annealing temperature (51°C), 1 min at 72°C and

10 mins at 72°C. PCR cycles were 28 and 35 for GAPDH and GR respectively. Following separation of the PCR products on ethidium bromide-stained agarose gel (1%), the intensity of the bands were quantified densitometrically using image analyser (Gel Doc 2000, Bio-Rad, Munich, Germany). Each band was normalised against the intensity obtained with the same cDNA using the glyceraldehyde-3-phosphate dehydrogenase (GAPDH)-specific primers which served as internal controls. The following primers were used: GAPDH (forw) 5'-AGTCGCTCTTCAAAGGGATG-3'; GAPDH (rev) 5'-TCAGTGTAGCCAAGGATACC-3'; GR (forw) 5'-C(T)C(T)TA(T/G)ACA(T/C)CCA(T/G)GTT(G)GCA(T/C)C(T)TGATG-3'; GR (rev) 5'-CT(G)CATT(C/G)GTC(G)ACA(G)AAC(T)TCC(T)TCA(T)-3'.

2.2.12.4 Primer design

New degenerate GR primers for RT-PCR were designed by comparative alignments of GR nucleotide sequences of different plants (Tab. 5) published in the National Center for Biotechnology Information (NCBI). Forward and reverse GR primers were designed from the conserved regions identified using Clone Manager Professional Suite software (Scientific & Educational Software, NC, USA). Primers were tested by performing PCR using cDNA reverse-transcribed from mRNA isolated from *C. demersum*. The PCR product was run on a 1% agarose gel by gel-electrophoresis.

Table 5. GenBank accession numbers of plant species used for designing degenerate GR primers

Plant species	Accession number
Triticum monococcum	AY364467
Pisum sativum	X98274
Zea mays	AJ006055
Nicotiana tabacum	X76533
Zinnia elegans	AB158513
Oryza sativa (holotype f)	AY136765

2.2.12.5 Purification of DNA from gel and sequencing

Band of expected size was excised using a scalpel disinfected with ethanol. DNA was purified from the gel using a purification kit (GFX PCR DNA and Gel Band Purification Kit, Amersham Biosciences) according to the manufacturer's protocol. In order to dissolve the gel, 10 µl of capture buffer was added to 10 mg of gel in a 1.5 ml Eppendorf tube and then placed in a Thermomixer and shaked intermittently for 10 minutes at 60°C. The dissolved gel sample was transferred to a GFX column and incubated at room temperature for 1 min. The sample was centrifuged at 13,400 rpm for 30 sec and the flow-through was discarded. 500 µl washing buffer was added to the column and centrifuged for 30 seconds at 13,400 rpm. The column was transferred to a 1.5 ml Eppendorf tube and eluted with 10 µl of DEPC- treated water. DNA concentration in the eluate was determined spectrophotometrically at 280 nm and thereafter forwarded for automated sequencing (AGOWA, Germany). The obtained sequence was verified for homology with GR nucleotide sequences of other plants published in NCBI by a Basic Local Alignment Search Tool (BLAST) using the software Clone Manager Professional Suite (Scientific & Educational Software, NC, USA).

2.2.13 Statistical Analysis

Data sets were checked for variance homogeneity (Levenes test) and normality (Shapirow-Wilk's test) (Zar, 1996). Analysis of Variance (ANOVA) was then performed to test for differences among treatments. One-way ANOVA was used throughout in dose-response experiments. In time-response experiments, two-factor (time and leaf extract) ANOVA was used. Where ANOVA indicated significant differences, planned post-hoc comparisons or contrasts were also performed using either Tukeys Honest Significant Difference (HSD) or Newmann-Keuls test. In cases where unequal number of samples was involved, Tukeys HSD for unequal 'n' test was used. All comparisons were made at the 5% probability level of significance. The statistical software STATISTICA (StatSoft, Inc. 2000) was used in all statistical analyses.

In *Lemna* bioassay experiments, a completely randomised design (CRD) was used with 4 replicates for each treatment and control. Data were also checked for homogeneity of variance and normality and were natural-log (ln) or square root-transformed when necessary to improve normality (Zar, 1996). Data were then subjected to one-way analysis of variance (ANOVA) followed by Tukeys HSD test. Dunnett's post-hoc test was used to determine lowest and no observed effect concentrations (LOEC and NOEC). The statistical software STATISTICA (StatSoft, Inc. 2000) was used. Day zero (0) data were neither used in statistical treatments nor in subsequent discussion; they were presented only to show the physiological status of *Lemna* plants at the time of exposure.

3. Results

3.1 Experiments with *Ceratophyllum demersum*

3.1.1 Experiment 1: Dose-response effects of *Phragmites australis* and *Quercus robur* leaf litter extracts on *C. demersum*

3.1.1.1 Dissolved organic carbon fractions in *P.australis* and *Q. robur* leaf extracts

Prior to exposure experiments, both leaf litter extracts were characterized by LC-OCD into different molecular weight fractions and aromaticity (Table 6). Extracts from *Q. robur* leaf litter contained higher amounts of dissolved organic carbon (DOC) fractions than *P. australis* leaf extracts. Humic-like substances (HS) make up 32% of the DOC fractions in *Q. robur* extracts and 29.1% in *P. australis* leaf extract. Aromaticity was by a factor of two higher in *Q. robur* leaf litter extract than in *P. australis* leaf litter extract. Conversely, the proportion of low molecular weight substances (LMWS) was higher in *P. australis* than in *Q. robur* extracts.

Table 6. Dissolved organic carbon fractions and aromaticity from *Quercus robur* (oak) and *Phragmites australis* (reed) leaf extracts

Plant species	Total DOC (mg L^{-1})	Aromaticity (L $mg^{-1}m^{-1}$)	HS (%)	HMWS (%)	LMWS (%)
Phragmites australis	404.3	2.2	29.1	8.6	62.3
Quercus robur	3845.6	4.0	32.2	26	41.8

HS: Humic-like substances, HMWS: High molecular weight substances, LMWS: Low molecular weight substances, DOC: Dissolved organic carbon.

3.1.1.2 Photosynthetic oxygen release in *C. demersum*

Direct exposure of the aquatic macrophyte, *C. demersum*, to leaf litter extracts of *P. australis* and *Q. robur* showed a clear adverse effect on photosynthetic oxygen production (Fig. 9). The effects were significant at all tested DOC concentrations in both leaf extracts.

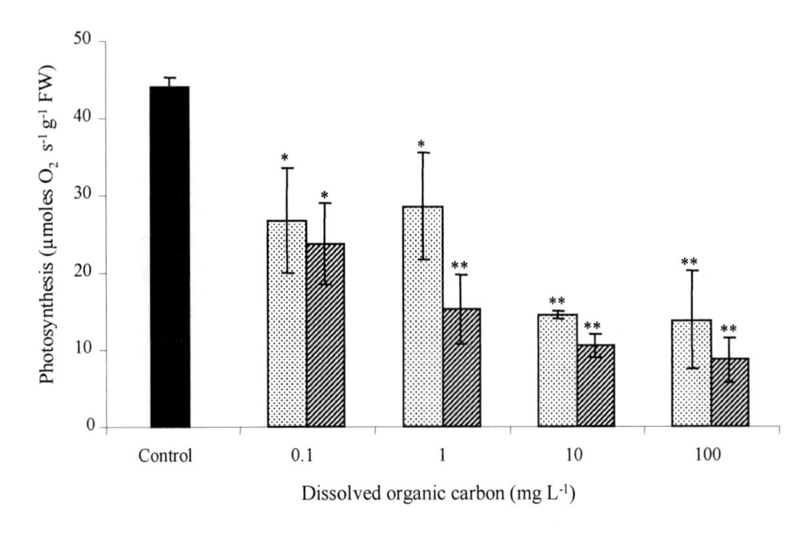

Figure 9. Dose-response photosynthetic oxygen production in *C. demersum* after 24 h exposure to different dissolved organic carbon concentrations from two plant leaf-litter extracts. Values are means (n = 5) ± standard deviation, * $p < 0.05$, ** $p < 0.01$ (Tukeys HSD test), [dotted pattern] *P. australis* extract, [hatched pattern] *Q. robur* extracts.

At the highest concentrations (10 and 100 mg L^{-1} DOC), very strong effects were observed with a highly significant ($p < 0.01$) reduction in the rate of photosynthetic oxygen release. Photosynthetic rate of *C. demersum* was more reduced in *Q. robur* than in *P. australis* extracts, especially at the 1 and 10 mg L^{-1} DOC concentrations, compared to control.

3.1.1.3 Chlorophyll pigments in *C. demersum*

During 24 h of exposure to *P. australis* and *Q. robur* extracts, *C. demersum* manifested no detectable effects on chlorophyll pigments. Both total chlorophyll (Fig. 10a) and chlorophyll a/b ratio (Fig. 10b) remained unchanged at all DOC concentrations tested, compared with control values.

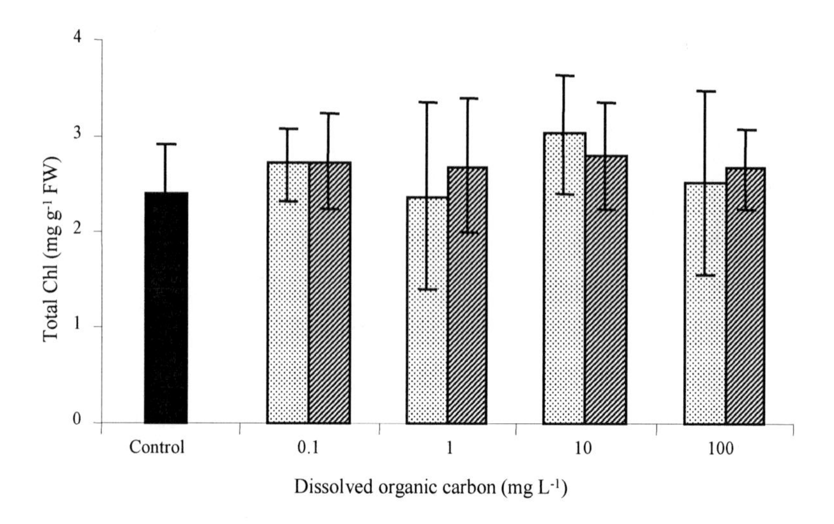

Figure 10a. Dose-response total chlorophyll in *C. demersum* after 24 h exposure to different dissolved organic carbon concentrations from two plant leaf-litter extracts. Values are means (n = 5) ± standard deviation, p > 0.05 (Tukeys HSD test), *P. australis* extract, *Q. robur* extracts.

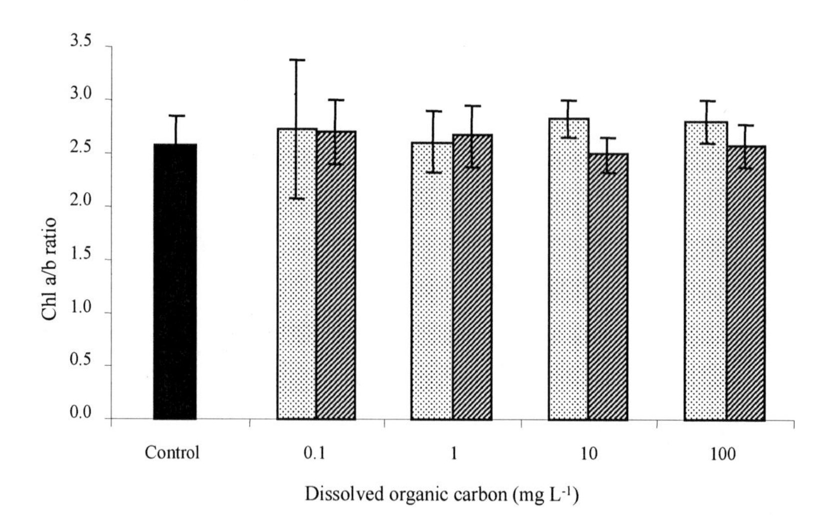

Figure 10b. Dose-response chlorophyll a/b ratio in *C. demersum* after 24 h exposure to different dissolved organic carbon concentrations from two plant leaf-litter extracts. Values are means (n = 5) ± standard deviation, p > 0.05 (Tukeys HSD test), *P. australis* extract, *Q. robur* extract.

3.1.1.4 Enzymatic activity response of microsomal glutathione S-transferase (mGST), assayed with the conjugating substrate 1-chloro-2,4-dintrobenzene (CDNB), in *C. demersum*

Microsomal glutathione S-transferase activity, measured with CDNB as substrate, was significantly elevated in exposed *C. demersum* compared with control (Fig. 11). In both *P. australis* and *Q. robur* extracts, the increase in activity was significant at all tested DOC concentrations.

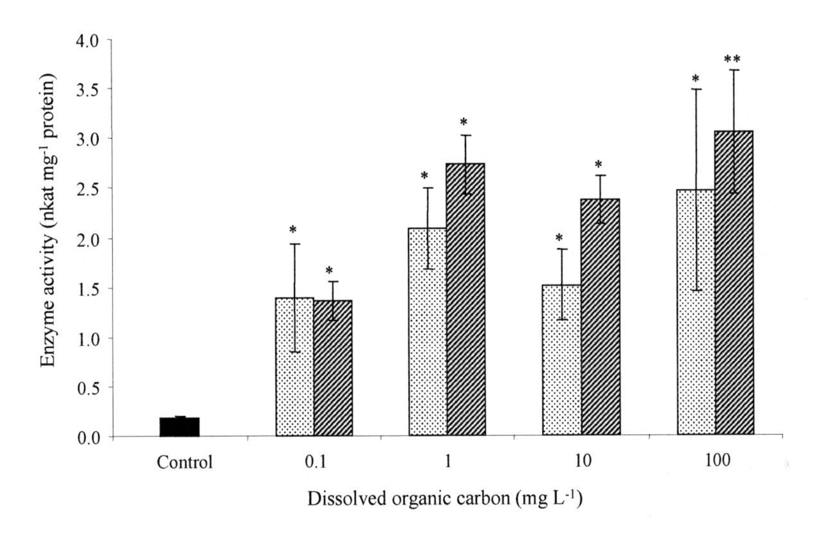

Figure 11. Dose-response of microsomal glutathione S-transferase, measured with CDNB, in *C. demersum* after 24 h exposure to different dissolved organic carbon concentrations from two plant leaf-litter extracts. Values are means (n = 5) ± standard deviation, * $p < 0.05$, ** $p < 0.01$ (Tukeys HSD test), *P. australis* extract, *Q. robur* extracts.

3.1.1.5 Enzymatic activity response of cytosolic glutathione S-transferase (cGST), assayed with the conjugating substrate CDNB, in *C. demersum*

Similar to mGST, the activity of cytosolic GST (cGST), assayed with CDNB as substrate, was clearly induced at all tested DOC concentrations from *P. australis* and *Q. robur* (Fig. 12). In *Q. robur* extracts, the increase in activity of the enzyme is dose-dependent, whereas in *P. australis* extract, enzyme activity tends to decrease at higher DOC levels.

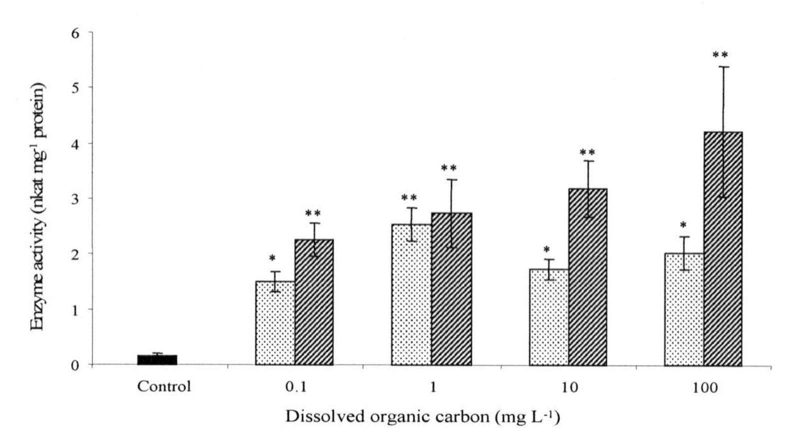

Figure 12. Dose-response of cytosolic glutathione S-transferase, measured with CDNB, in *C. demersum* after 24 h exposure to different dissolved organic carbon concentrations from two plant leaf-litter extracts. Values are means (n = 5) ± standard deviation, * $p < 0.05$, ** $p < 0.01$ (Tukeys HSD test), *P. australis* extract, *Q. robur* extracts.

3.1.1.6 Enzymatic activity response of cytosolic glutathione S-transferase (cGST), assayed with 4-hydroxynonenal as conjugating substrate, in *C. demersum*

A concentration-dependent significant increase in soluble cGST activity, using 4-hydroxynonenal as substrate, was detected in *C. demersum* exposed to *Q. robur extract* (Fig. 13).

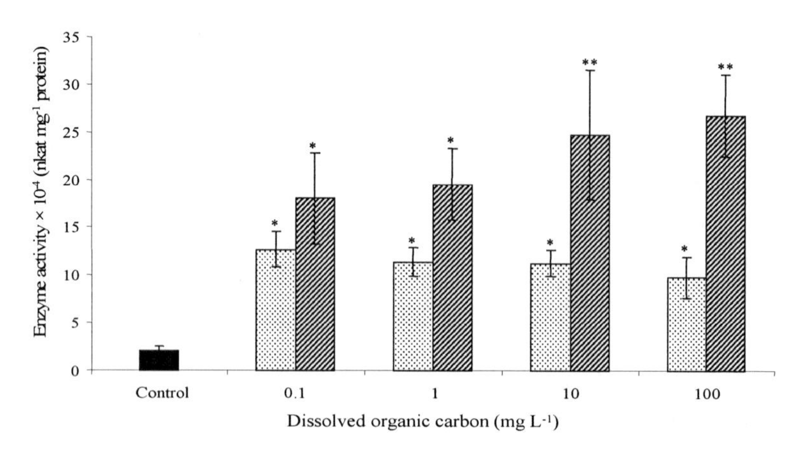

Figure 13. Dose-response of cytosolic glutathione S-transferase, measured with 4-hydroxynonenal, in *C. demersum* after 24 h exposure to different dissolved organic carbon concentrations from two plant leaf-litter extracts. Values are means (n = 5) ± standard deviation, * $p < 0.05$, ** $p < 0.01$ (Tukeys HSD test), *P. australis* extract, *Q. robur* extracts.

In *C. demersum* exposed to *P. australis* extract, the increase in cGST activity was significant at all DOC concentrations tested but not in a concentration-dependent manner. At higher *P. australis*-DOC concentrations (10 and 100 mg L^{-1}), the activity of cGST decreased slightly.

3.1.1.7 Enzymatic activity response of glutathione reductase (GR) in *C. demersum*

Glutathione reductase activity in *C. demersum* exposed to *P. australis* and *Q. robur* extracts increased significantly compared to untreated control, showing a tendency to decrease at the highest DOC concentration (Fig. 14).

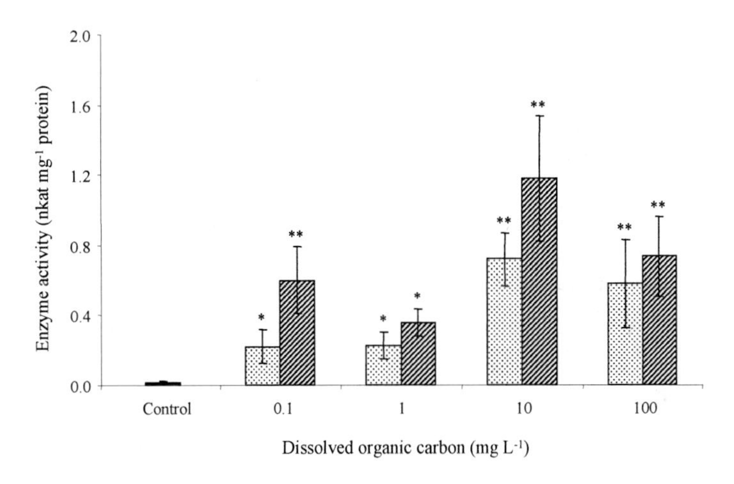

Figure 14. Dose-response of glutathione reductase in *C. demersum* after 24 h exposure to different dissolved organic carbon concentrations from two plant leaf-litter extracts. Values are means (n = 5) ± standard deviation, * $p < 0.05$, ** $p < 0.01$ (Tukeys HSD test), *P. australis* extract, *Q. robur* extracts.

3.1.1.8 Enzymatic activity response of guaiacol peroxidase (POD) in *C. demersum*

Activity of guaiacol peroxidase in *C. demersum* increased significantly at all doses compared with control and in both *P. australis* and *Q. robur* extracts (Fig. 15). Elevations in POD activity were higher in *C. demersum* exposed to *Q. robur* than that exposed to *P. australis* extract. Highest elevation compared to control was by a factor of 12 in *P. australis* exposure and by a factor of 47 in *Q. robur* exposure. *C. demersum* exposed to *P. australis* showed a gradual decrease in POD activity as the concentration increased from 0.1 to 100 mg L^{-1} DOC.

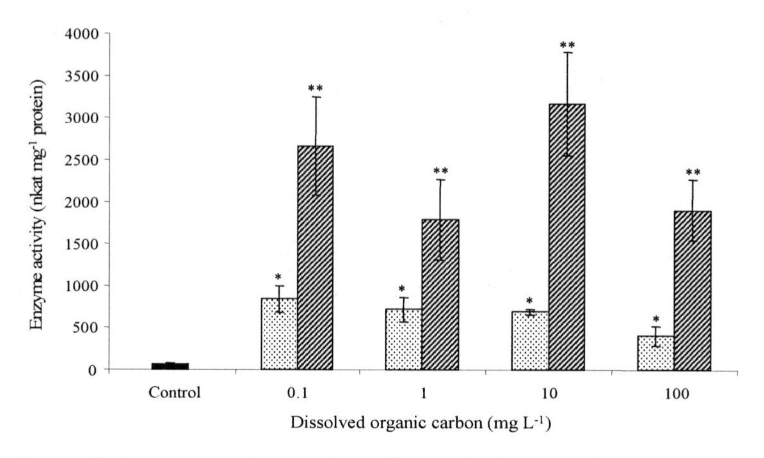

Figure 15. Dose-response of guaiacol peroxidase in *C. demersum* after 24 h exposure to different dissolved organic carbon concentrations from two plant leaf-litter extracts. Values are means (n = 5) ± standard deviation, * $p < 0.05$, ** $p < 0.01$ (Tukeys HSD test), *P. australis* extract, *Q. robur* extracts.

3.1.1.9 Lipid peroxidation (LPO) in the aquatic macrophyte *C. demersum*

No significant differences ($p > 0.05$) in concentration of malondialdehyde (MDA), an index of lipid peroxidation, was detected between control and *C. demersum* exposed to both *Q. robur* and *P. australis* extracts at all DOC concentrations tested (Fig. 16).

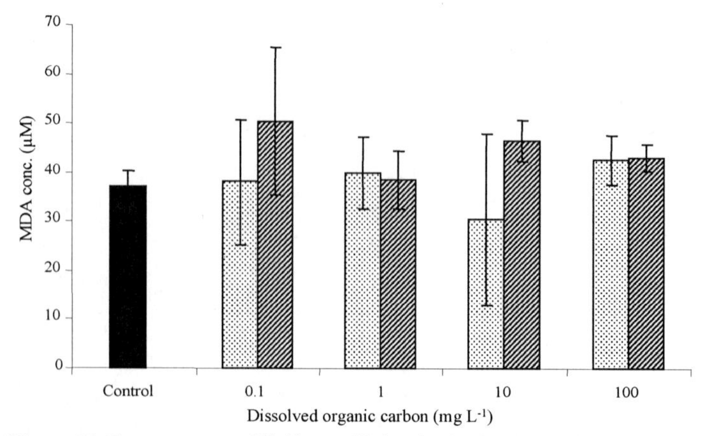

Figure 16. Dose-response of lipid peroxidation in *C. demersum* after 24 h exposure to different dissolved organic carbon concentrations from two plant leaf-litter extracts. Values are means (n = 5) ± standard deviation, $p > 0.05$ (Tukeys HSD test), *P. australis* extract, *Q. robur* extracts.

3.1.2 Experiment 2: Acclimation experiments with *C. demersum* exposed to *P. australis* and *Q. robur* extracts

3.1.2.1 Photosynthetic response on time-dependent exposure of *C. demersum* to *P. australis* and *Q. robur* extracts

A different pattern of inhibitory response to photosynthetic oxygen production in *C. demersum* due to exposure to the two leaf extracts (*Phragmites australis* and *Quercus robur* extracts) was observed. Significant ($p < 0.05$) photosynthetic inhibition was observed after 4 h of exposure to *P. australis* extracts with a maximum reduction after 24 h (Fig. 17). On the other hand, photosynthetic oxygen production was significantly ($p < 0.05$) reduced after 12 h of exposure to *Q. robur* leaf extract. Similar to *P. australis* leaf extract, highest photosynthetic reduction was attained after 24 h of exposure to *Q. robur* extract. This photosynthetic inhibition by *Q. robur* extract persisted till 1 week (168 h). In contrast, a resumption of photosynthetic recovery was observed after 48 h of exposure to *P. australis* extract with complete recovery during one week (168 h) of exposure.

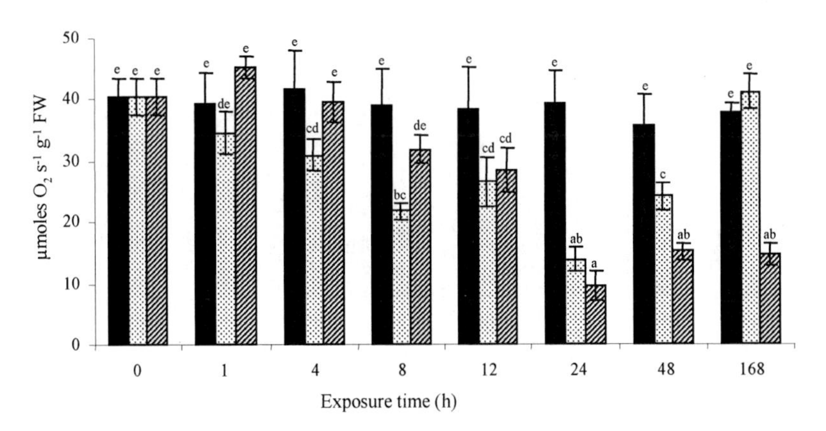

Figure 17. Time-response of photosynthetic oxygen release in *C. demersum* after exposure to 10 mg L^{-1} DOC from [dotted] *P. australis* and [hatched] *Q. robur* extracts and [black] control. Values with the same letter are not significantly different (Tukeys HSD test), n = 4.

3.1.2.2 Enzymatic response of microsomal glutathione S-transferase (mGST), measured with CDNB, on time-dependent exposure of *C. demersum* to *P. australis* and *Q. robur* extracts

Both leaf extracts induced significant ($p < 0.05$) increases in the microsomal GST (mGST) activity after 1 h of exposure and reached maximum levels after 24 h (Fig. 18). Thereafter (48 h), mGST levels in both extracts decreased and subsequently (at 168 h) returned to control values for *C. demersum* in *P. australis* extracts or remained elevated for *C. demersum* in *Q. robur* extract.

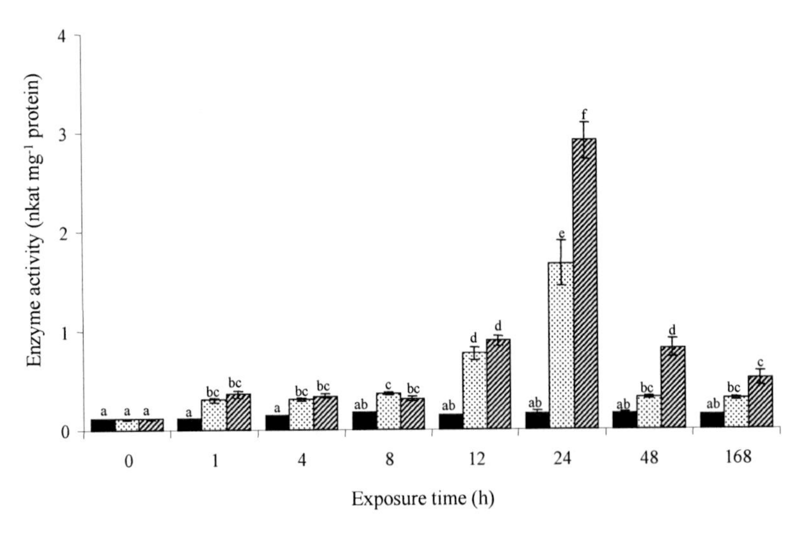

Figure 18. Time-response of microsomal glutathione S-transferase (mGST), measured with CDNB, in *C. demersum* after exposure to 10 mg L^{-1} DOC from *P. australis* and *Q. robur* extracts and control. Values with the same letter are not significantly different (Tukeys HSD test), n = 4.

3.1.2.3 Enzymatic response of cytosolic glutathione S-transferase (cGST), measured with CDNB, on time-dependent exposure of *C. demersum* to *P. australis* and *Q. robur* extracts

Soluble (cytosolic) glutathione S-transferase (cGST) activity in *C. demersum* exposed to *P. australis* extract was significantly ($p < 0.05$) induced at 8 h when considering 1-chloro-2,4-dinitrobenzene (CDNB) as the substrate (Fig. 19). On the other hand, significant ($p < 0.05$) induction in cGST (CDNB substrate) activity occurred after 1 h exposure to *Q. robur* leaf extract; further increase was observed only after 24 h. In both leaf extracts, highest cGST activity (CDNB substrate) in *C. demersum* was observed at 24 h. After 48 h exposure to both extracts, significant decrease in cGST (CDNB substrate) compared with the one noted at 24 h was observed. No further decrease in cGST (CDNB substrate) occurred at 168 h following exposure to *Q. robur* extract. cGST (CDNB substrate) activity returned to control values after 168 h exposure to *P. australis* extract.

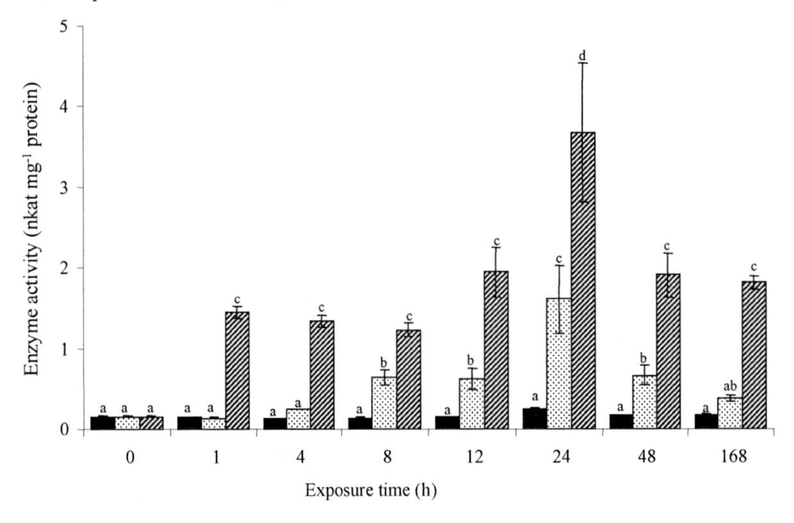

Figure 19. Time-response of cytosolic glutathione S-transferase (cGST), measured with CDNB, in *C. demersum* after exposure to 10 mg L^{-1} DOC from *P. australis* and *Q. robur* extracts and control. Values with the same letter are not significantly different (Tukeys HSD test), n = 4.

3.1.2.4 Enzymatic response of cytosolic glutathione S-transferase (cGST), measured with 4-hydroxynonenal (4-HNE), on time-dependent exposure of *C. demersum* to *P. australis* and *Q. robur* extracts

Considering 4-hydroxynonenal (4-HNE) as substrate, cGST activity in *C. demersum* exposed to *P. australis* extract was elevated at 4 h and reached a maximum activity at 12 h (Fig. 20). Significant reduction in cGST (4-HNE substrate) activity was detected at 48 h when compared with the ones noted at 12 and 24 h in *P. australis* extract. After 168 h, cGST (4-HNE substrate) activity returned to control values for *C. demersum* exposed to *P. australis* extracts. In contrast, significant increase in cGST (4-HNE substrate) activity was induced after 1 h exposure of *C. demersum* to *Q. robur* extract. Highest cGST (4-HNE substrate) activity was attained after 24 h in *Q. robur* extracts, followed by a reduction in activity that did not reach control values after 168 h.

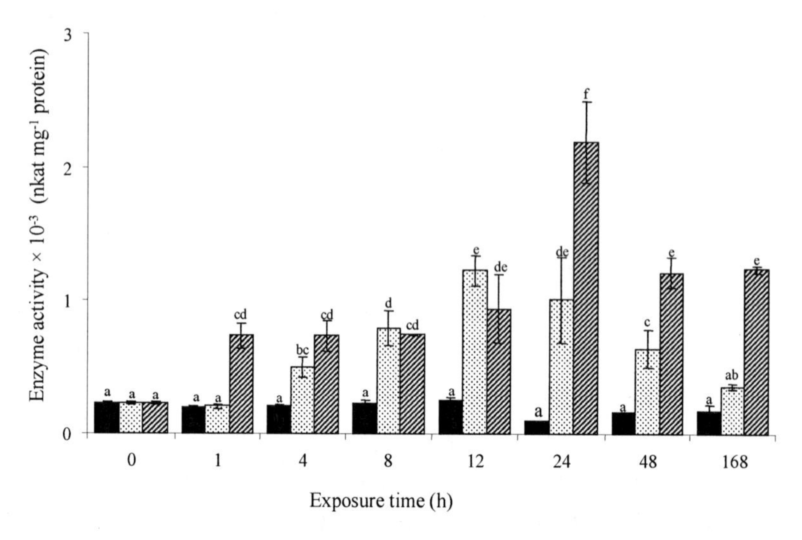

Figure 20. Time-response of cytosolic glutathione S-transferase (cGST), measured with 4-HNE, in *C. demersum* after exposure to 10 mg L^{-1} DOC from *P. australis* and *Q. robur* extracts and control. Values with the same letter are not significantly different (Tukeys HSD test), n = 4.

3.1.2.5 Enzymatic response of guaiacol peroxidase (POD) on time-dependent exposure of *C. demersum* to *P. australis* and *Q. robur* extracts

Peroxidase (POD) activity in *C. demersum* was significantly ($p < 0.05$) elevated after 1 and 8 h exposures to *Q. robur* and *P. australis* extracts respectively (Fig. 21). After 24 h exposure, both extracts induced the highest POD activity. Subsequently, POD activity decreased and after 168 h reached control values for *C. demersum* exposed to *P. australis* extracts. In *Q. robur* extracts, POD level remained significantly ($p < 0.05$) higher than control values till the end of the experiment.

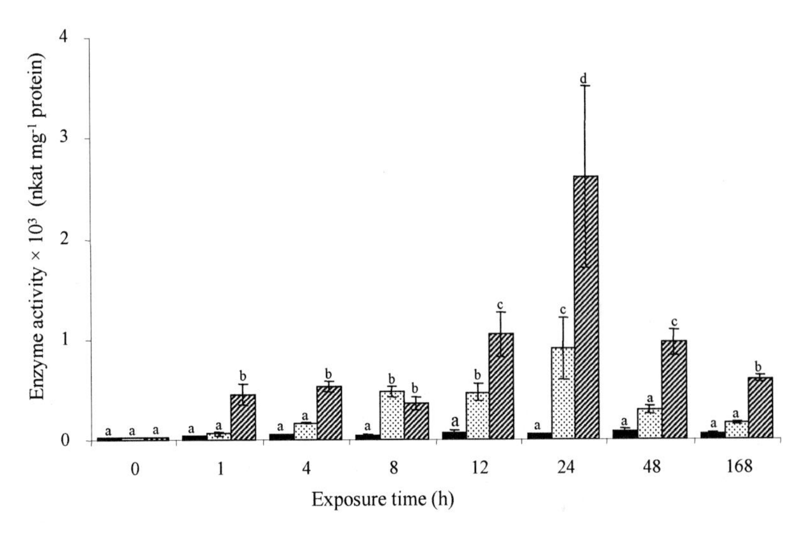

Figure 21. Time-response of guaiacol peroxidase (POD) in *C. demersum* after exposure to 10 mg L^{-1} DOC from *P. australis* and *Q. robur* extracts and control. Values with the same letter are not significantly different (Tukeys HSD test), n = 4.

3.1.2.6 Enzymatic response of glutathione peroxidase (GPx) on time-dependent exposure of *C. demersum* to *P. australis* and *Q. robur* extracts

Exposure of *C. demersum* to the two leaf extracts resulted in significant ($p < 0.05$) induction in glutathione peroxidase (GPx) activity after 4 h (*Q. robur* extract) or 8 h (*P. australis* extract) (Fig. 22). The effect on GPx activity in both extracts was sustained until 24 h, although no significant ($p > 0.05$) effect was observed in *P. australis* extract at 12 h. At the end of the experiment (168 h), GPx returned to control values for *C. demersum* exposed to *P. australis* extract but remained elevated for *C. demersum* exposed to *Q. robur* extract.

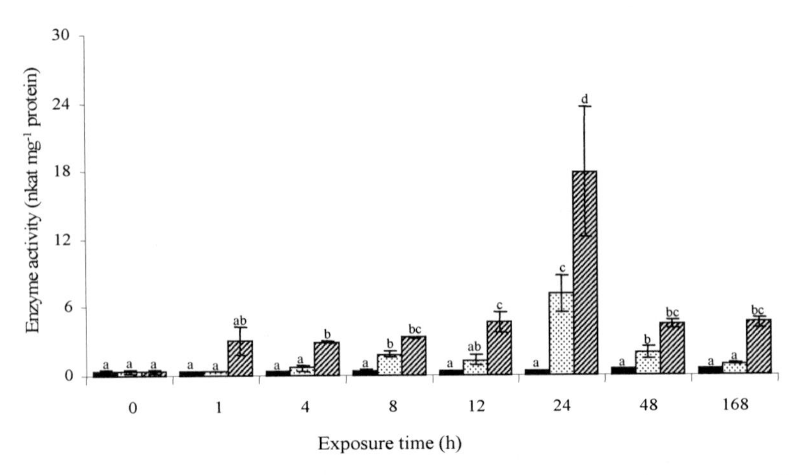

Figure 22. Time-response of glutathione peroxidase (GPx) in *C. demersum* after exposure to 10 mg L^{-1} DOC from *P. australis* and *Q. robur* extracts and control. Values with the same letter are not significantly different (Tukeys HSD test), n = 4.

3.1.2.7 Enzymatic response of glutathione reductase (GR) on time-dependent exposure of *C. demersum* to *P. australis* and *Q. robur* extracts

Both extracts did not show any significant effect on the glutathione reductase (GR) activity in *C. demersum* until after 8 h (*P. australis* extract) or 24 h (*Q. robur*) exposure (Fig. 23).

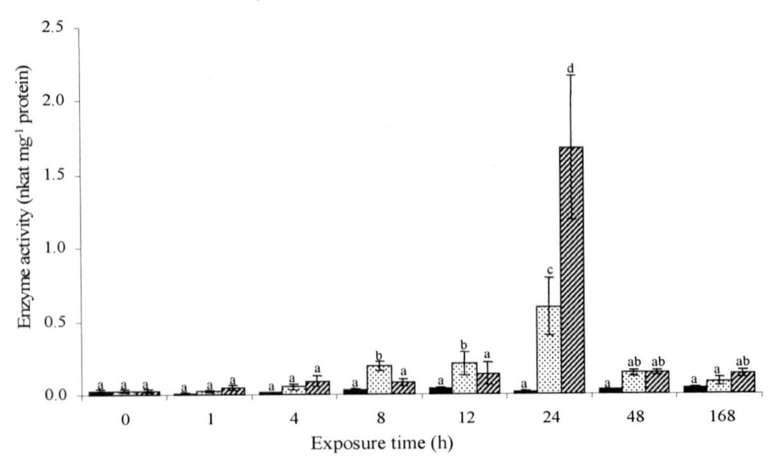

Figure 23. Time-response of glutathione reductase (GR) in *C. demersum* after exposure to 10 mg L^{-1} DOC from *P. australis* and *Q. robur* extracts and control. Values with the same letter are not significantly different (Tukeys HSD test), n = 4.

In both extracts, the effect was not sustained beyond 24 h as the activities returned to control levels.

3.1.3 Experiment 3: Effects of *P. australis* and *Q. robur* leaf litter extracts on intracellular hydrogen peroxide content, glutathione content and glutathione reductase gene expression in *C. demersum*.

3.1.3.1 Hydrogen peroxide content in the aquatic macrophyte *C. demersum*

C. demersum exhibited a dose-dependent increase in H_2O_2 content when exposed to *Q. robur* leaf extracts for 24 h (Fig. 24). Similar to *Q. robur* extract, *C. demersum* exposed to *P. australis* extract showed significant ($p < 0.01$) increases in H_2O_2 content compared with control in all DOC concentrations tested. A two-fold increase in H_2O_2 concentration was observed at the highest DOC concentration in *Q. robur* extract. In both leaf extracts, H_2O_2 levels tend to plateau at high DOC concentrations (10 and 100 mg L^{-1}).

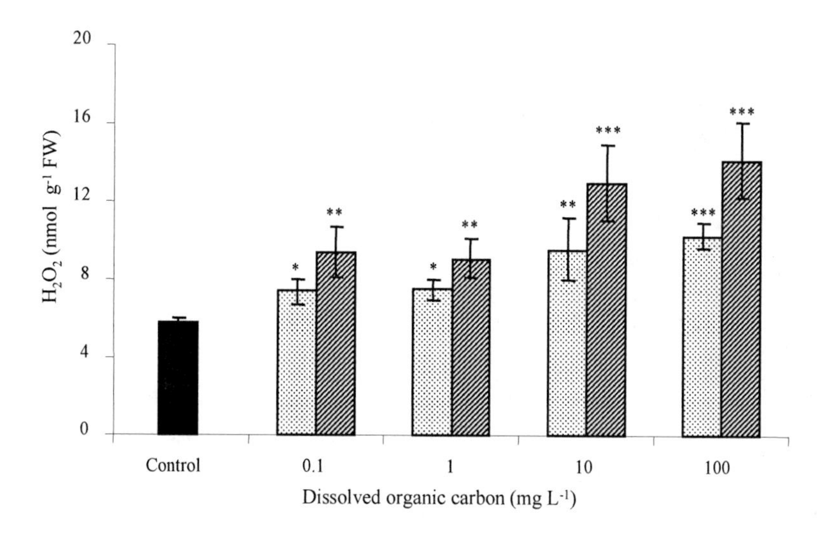

Figure 24. Dose-response of intracellular hydrogen peroxide in *C. demersum* after 24 h exposure to different dissolved organic carbon concentrations from two plant leaf-litter extracts. Values are means (n = 4) ± standard deviation, * $p < 0.05$, ** $p < 0.01$, *** $p < 0.001$ (Newman-Keuls test), *P. australis* extract, *Q. robur* extracts.

3.1.3.2 Total glutathione contents in the aquatic macrophyte *C. demersum*

In parallel to increased H_2O_2 content, total glutathione content was significantly elevated for *C. demesum* exposed to *Q. robur* extracts except at 100 mg L^{-1} (Fig. 25a). In *C. demersum* exposed to *P. australis* extracts, no significant ($p > 0.05$) increase in total glutathione content was observed compared with control level at the lowest DOC concentration (0.1 mg L^{-1}), but significant increase occurred at 1, 10 and 100 mg L^{-1} DOC.

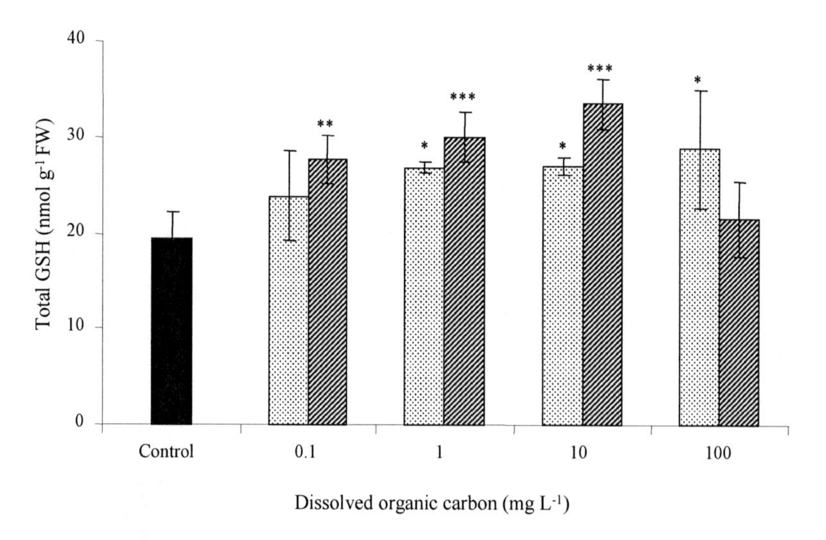

Figure 25a. Dose-response of total glutathione content in *C. demersum* after 24 h exposure to different dissolved organic carbon concentrations from two plant leaf-litter extracts. Values are means (n = 4) ± standard deviation, * $p < 0.05$, ** $p < 0.01$, *** $p < 0.001$ (Newman-Keuls test), [dotted] *P. australis* extract, [hatched] *Q. robur* extracts.

Time response experiments were conducted using 10 mg L^{-1} DOC from *P. australis* and *Q. robur* leaf litter extracts and results for total glutathione are shown in Figure 25b. In both leaf extracts, *C. demersum* exhibited similar total glutathione levels throughout the experiment except at 12 and 24 h (*P. australis*) and 24 h (*Q. robur*) where significantly higher levels were noted (Fig. 25b).

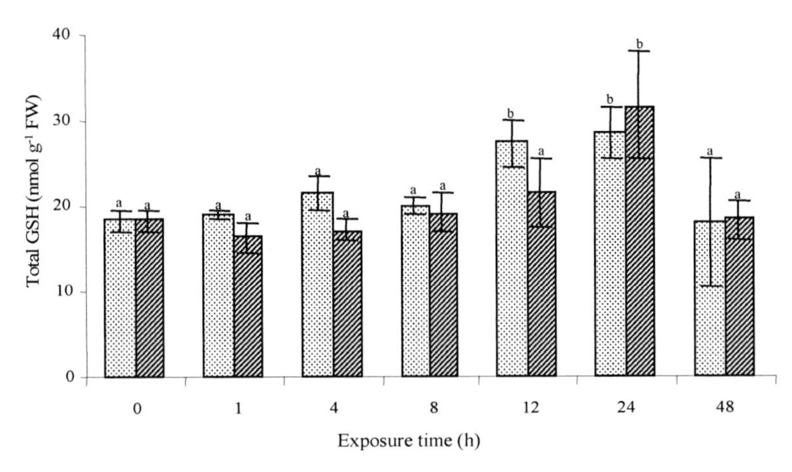

Figure 25b. Time-response of total glutathione content in *C. demersum* after exposure to 10 mg L^{-1} DOC from *P. australis* and *Q. robur* extracts. Values with the same letter are not significantly different (Newman-Keuls test), n = 4.

3.1.3.3 Reduced to oxidised (GSH/GSSG) glutathione ratio in the aquatic macrophyte *C. demersum*

When exposed to *Q. robur* extracts, *C. demersum* exhibited a dose-dependent decrease in GSH/GSSG ratio (Fig. 26a).

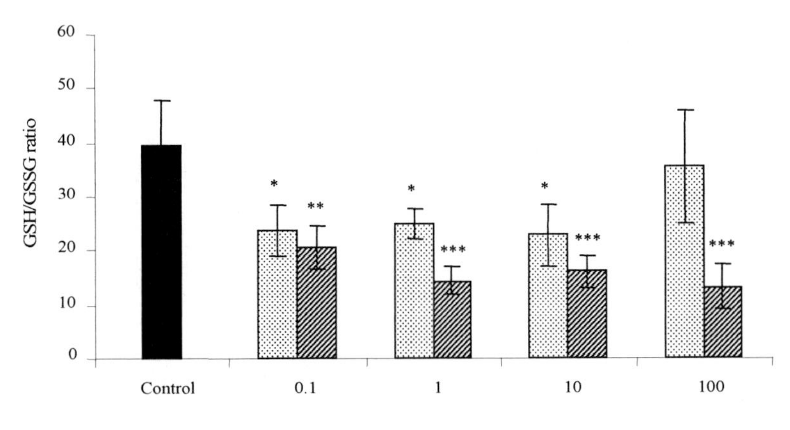

Figure 26a. Dose-response of GSH/GSSG ratio in *C. demersum* after 24 h exposure to different dissolved organic carbon concentrations from two plant leaf-litter extracts. Values are means (n = 4) ± standard deviation, * $p < 0.05$, ** $p < 0.01$, *** $p < 0.001$ (Newman-Keuls test), *P. australis* extract, *Q. robur* extracts.

By contrast, a non dose-dependent decline in GSH/GSSG ratio was observed in *C. demersum* exposed to *P. australis* extract. At the highest DOC concentration (100 mg/L) in *P. australis* extract, GSH/GSSG ratio was not significantly ($p > 0.05$) different from control value.

In time response of *C. demersum* exposed to *P. australis* extract, GSH/GSSG ratio declined significantly at 1, 4, 8 and 12 h compared with the ratio noted at 0 h (Fig. 26b). After 24 and 48 h, GSH/GSSG ratios in *P. australis* extract increased and reached levels comparable to the one noted at 0 h. In *Q. robur* extracts, GSH/GSSG ratio was significantly lower at all time points compared with the ratio observed at 0 h. Lower GSH/GSSG ratios were observed in *Q. robur* (except 0 h) than in *P. australis* extract throughout the experiment.

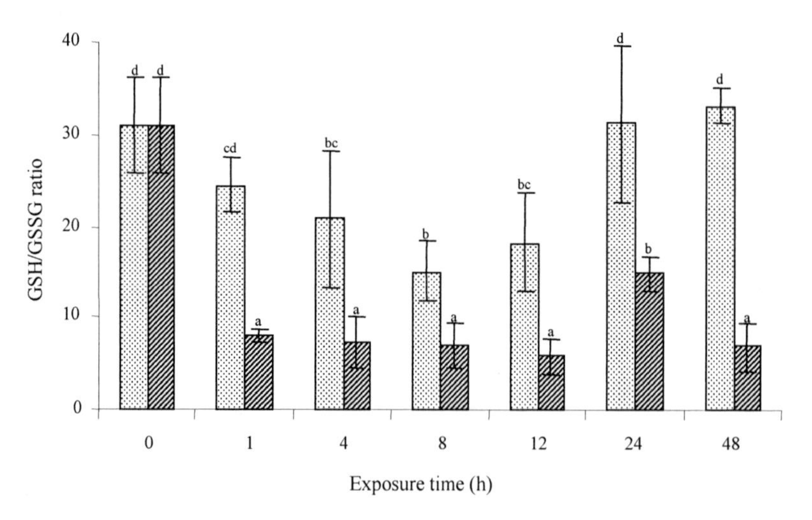

Figure 26b. Time-response of GSH/GSSG ratio in *C. demersum* after exposure to 10 mg L^{-1} DOC from *P. australis* and *Q. robur* extracts. Values with the same letter are not significantly different (Newman-Keuls test), n = 4.

3.1.3.4 Glutathione reductase expression in the aquatic macrophyte *C. demersum*

The partial glutathione reductase (GR) protein sequence (GenBank accession no. EF495224) isolated from *C. demersum* is shown in Figure 27. The sequence consists of 359 nucleotide base pairs and 119 translated amino acids.

```
  1  TGGCCTTGAT GGAAGGGACA AGTTTTTCAA AAACTGTATT TGGAGGACAA CCTACTAAAC
  1     A  L   M  E  G  T   S  F  S   K  T  V   F  G  G  Q   P  T  K
 61  CTGATTACAA TTATGTTCCT TGTGCAGTTT TTTGTGTCCC ACCCCTCTCA GTGGTTGGCT
 20  P  D  Y   N  Y  V  P   C  A  V   F  C  V   P  P  L  S   V  V  G
121  TCAGTGAGCA GCAGGCCATA GATCAAGCAA AGGGTGATAT TCTGATTTTC ACATCAACAT
 40  F  S  E   Q  Q  A  I   D  Q  A   K  G  D   I  L  I  F   T  S  T
181  TCAATCCAAT GAAGAACACC ATCTCAGGGA GGCAAGAAAA GTCAATTATG AAACTGGTCG
 60  F  N  P   M  K  N  T   I  S  G   R  Q  E   K  S  I  M   K  L  V
241  TTGATGCTGA AACAGACAAA GTTCTTGGGG CATCCATGTG TGGTCCAGAT GCACCTGAAA
 80  V  D  A   E  T  D  K   V  L  G   A  S  M   C  G  P  D   A  P  E
301  TCATGCAGGG TCTTGCAATC GCTATCAAAT GTGGGGCAAC TAAAGCACAA TTGACAACA
100  I  M  Q   G  L  A  I   A  I  K   C  G  A   T  K  A  Q   L  T  T
```

Figure 27. Partial sequence of *C. demersum* glutathione reductase (GR) gene (GenBank accession no. EF495224). Translated amino acids are below the second nucleotide of each codon. Sequence and translation was compiled using Clone Manager Professional Suite software (Scientific & Educational Software, NC, USA). Numbers on the left represent nucleotide (regular numbers) and amino acid (bold numbers) positions.

The isolated protein sequence of *C. demersum* showed high homology with GR protein sequence of other plant species (Table 7). The highest (81%) and lowest (71%) similarity was with the plant species *Rheum australe* and *Arabidopsis thaliana*, respectively.

Table 7. Percentage similarity of GR protein sequence isolated from *C. demersum* compared by BLAST with GR protein sequence of other plants in GenBank.

Plant Species	Accession no.	% similarity
Brassica oleracea	BAD14936	73
Brassica rapa	AAC49980	73
Arabidopsis thaliana	AAK64087	71
Mesembryanthemum crystallinum	CAC13956	77
Pisum sativum	CAA66924	79
Rheum australe	ABI35910	81
Vigna unguilata	ABB89042	75
Zinnia elegans	BAD27393	78

P. australis extract did not result in any significant regulation of GR expression in *C. demersum* at the DOC levels tested within 24 h (Fig. 28a). *Q. robur* extract on the other hand induced a significant upregulation of GR expression at 10 mg L^{-1} DOC. At the highest tested DOC level (100 mg L^{-1}) in *Q robur* extract, GR expression in *C. demersum* was not significantly elevated after 24 h of exposure.

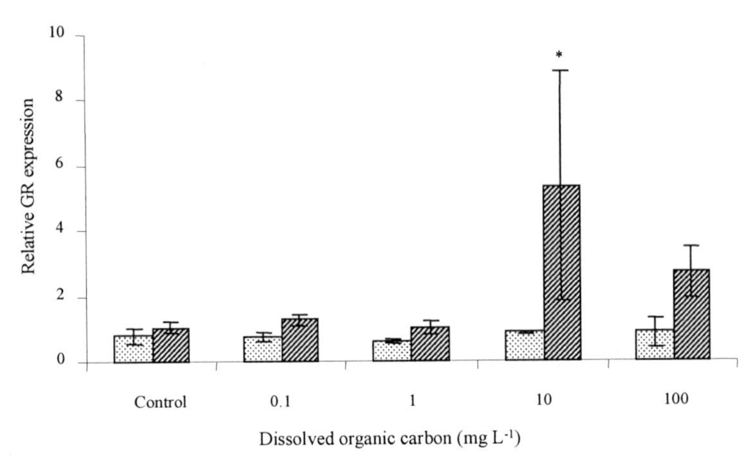

Figure 28a. Dose-response of glutathione reductase expression in *C. demersum* after 24 h exposure to different dissolved organic carbon concentrations from two plant leaf-litter extracts. Values are means (n = 4) ± standard deviation, * $p < 0.05$ (Newman-Keuls test), *P. australis* extract, *Q. robur* extracts.

Time response experiments on GR expression in *C. demersum* were conducted with 10 mg L^{-1} DOC of both leaf extracts and results are presented in Fig. 28b. When exposed to *P. australis* extract, *C. demersum* showed no significant change in GR expression at all time points except at 4 h where the expression was significantly lower compared to the ones observed at other time points.

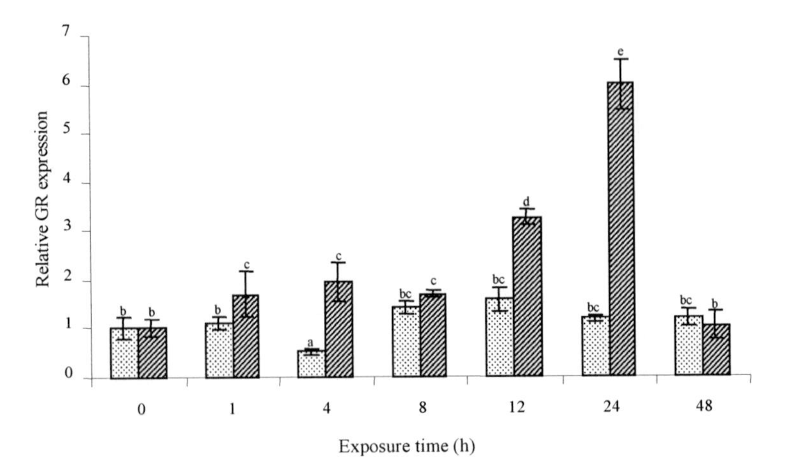

Figure 28b. Time-response of glutathione reductase expression in *C. demersum* after exposure to 10 mg L^{-1} DOC from *P. australis* and *Q. robur* extracts. Values with the same letter are not significantly different (Newman-Keuls test), n = 4.

For *C. demersum* exposed to *Q. robur* extract, GR expression was significantly higher at 1, 4, 8, 12 and 24 h compared with expression level noted at 0 h. After 48 h in *Q. robur* extract, the expression of GR returned to the level observed at 0 h. GR expression was in most cases higher (except at 0, 8 and 48 h) in *Q. robur* than in *P. australis* extract.

3.2 Experiments with *Lemna minor*

3.2.1 Experiment 4: Dose-response effects of *P. australis* and *Q. robur* leaf extracts on *L. minor*

3.2.1.1 Intracellular hydrogen peroxide content in the aquatic macrophyte *L. minor*

Exposure of *L. minor* to aqueous extracts from *Q. robur* resulted in dose-dependent increase in hydrogen peroxide contents (Fig. 29). For *L. minor* exposed to *P.australis* extract, a significant increase in hydrogen peroxide content was only observed at higher (10 and 100 mg L^{-1}) DOC concentrations.

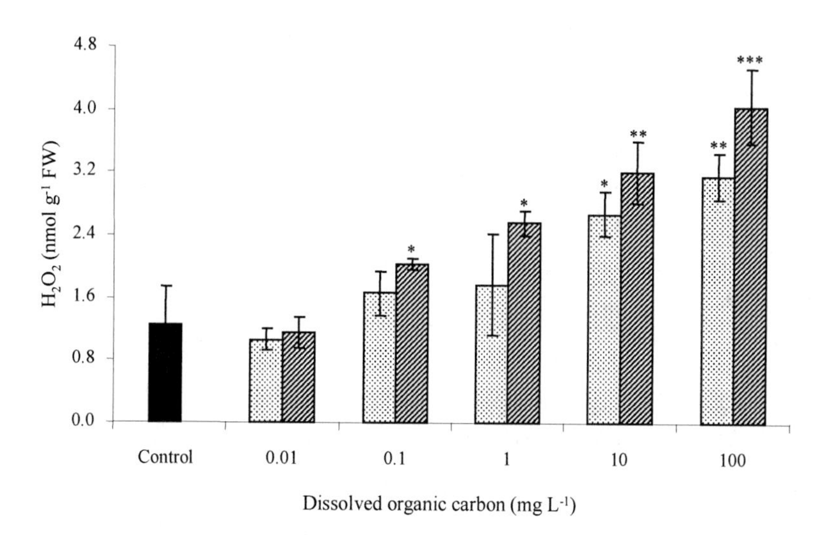

Figure 29. Dose-response of intracellular hydrogen peroxide in *L. minor* after 24 h exposure to different dissolved organic carbon concentrations from two plant leaf-litter extracts. Values are means ($n = 4$) ± standard deviation, * $p < 0.05$, ** $p < 0.01$, *** $p < 0.001$ (Tukeys HSD test), *P. australis* extract, *Q. robur* extract.

3.2.1.2 Total glutathione content in the floating aquatic macrophyte *L. minor*

Total glutathione levels in *L. minor* showed significant decrease when exposed to DOC concentrations of 100 mg L^{-1} (*Q. robur*) and 10 and 100 mg L^{-1} (*P. australis*) (Fig 30). Lower DOC concentrations (0.01 – 1 mg L^{-1}) in both extracts did not significantly alter the level of total glutathione in comparison with control levels.

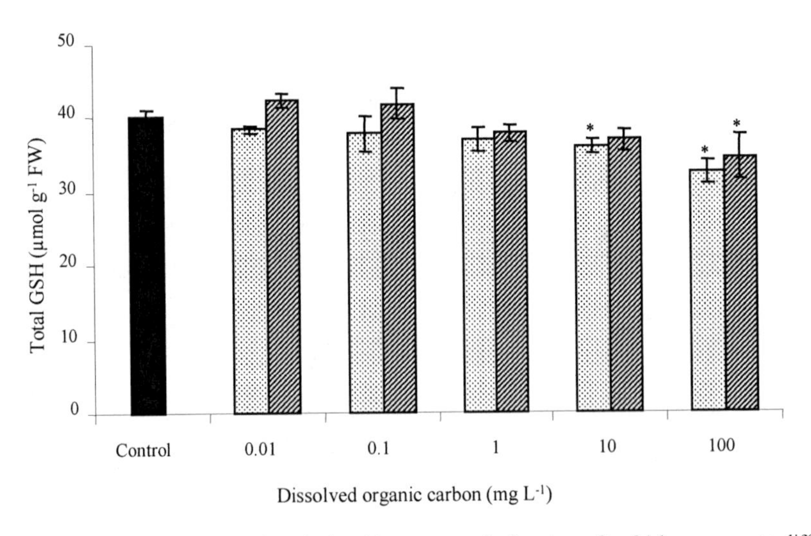

Figure 30. Dose-response of total glutathione content in *L. minor* after 24 h exposure to different dissolved organic carbon concentrations from two plant leaf-litter extracts. Values are means (n = 4) ± standard deviation, * $p < 0.05$ (Tukeys HSD test), *P. australis* extract, *Q. robur* extracts.

3.2.1.3 Reduced to oxidised (GSH/GSSG) glutathione ratio in the floating aquatic macrophyte *L. minor*

The ratio of reduced to oxidized (GSH/GSSG) glutathione in treated *L. minor* declined significantly at higher DOC levels in both extracts (Fig. 31). Only *Q. robur* extract caused an increase in GSH/GSSG ratio at 0.1 mg L^{-1}.

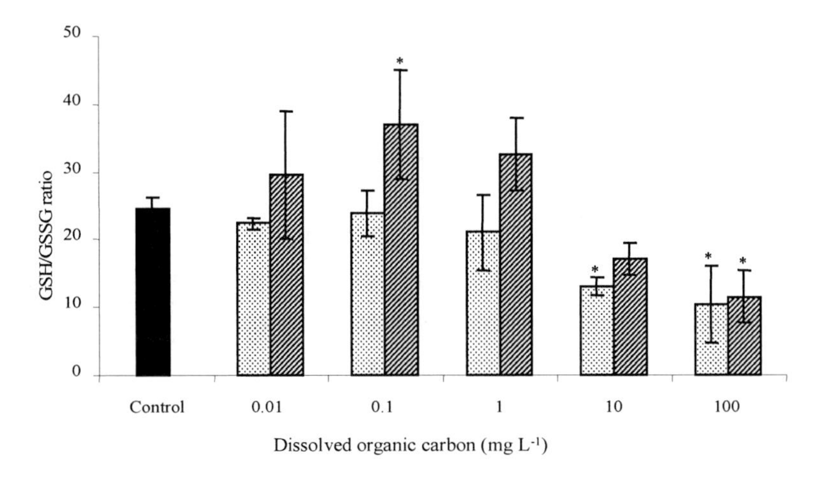

Figure 31. Dose-response of GSH/GSSG ratio in *L. minor* after 24 h exposure to different dissolved organic carbon concentrations from two plant leaf-litter extracts. Values are means (n = 4) ± standard deviation, * $p < 0.05$ (Tukeys HSD test), ▒ *P. australis* extract, ▨ *Q. robur* extracts.

3.2.1.4 Enzymatic activity response of superoxide dismutase in the floating aquatic macrophyte *L. minor*

In *L. minor* exposed to both *P. australis* and *Q. robur* extracts, SOD activity was induced at 1, 10 and 100 mg L^{-1} DOC (Fig. 32).

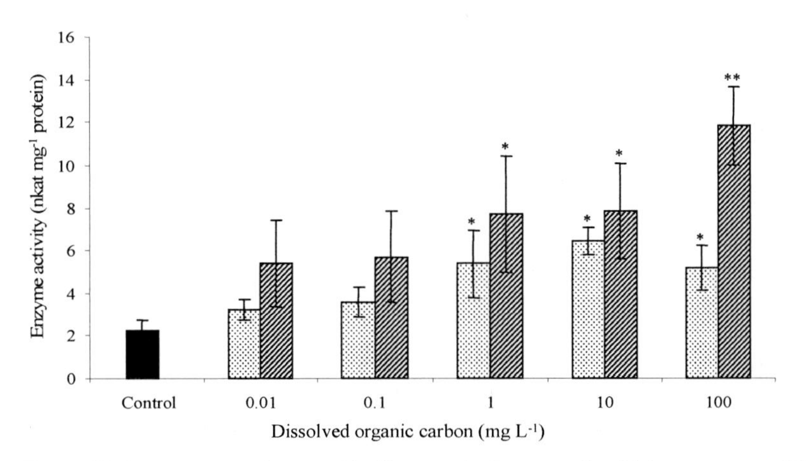

Figure 32. Dose-response of superoxide dismutase in *L. minor* after 24 h exposure to different dissolved organic carbon concentrations from two plant leaf-litter extracts. Values are means (n = 4) ± standard deviation, * $p < 0.05$, ** $p < 0.01$ (Tukeys HSD test), ▒ *P. australis* extract, ▨ *Q. robur* extract

At lower DOC concentrations, SOD activity in exposed *L. minor* was not statistically different from control values.

3.2.1.5 Enzymatic activity response of guaiacol peroxidase in *L. minor*

A dose-dependent increase in POD activity was observed in *L. minor* after 24 h exposure to *P. australis* extract at all tested DOC concentrations (Fig. 33). Similarly, *Q. robur* extract induced a dose-dependent increase (except at 0.01 mg L^{-1}) in POD activity which leveled off at high DOC concentration.

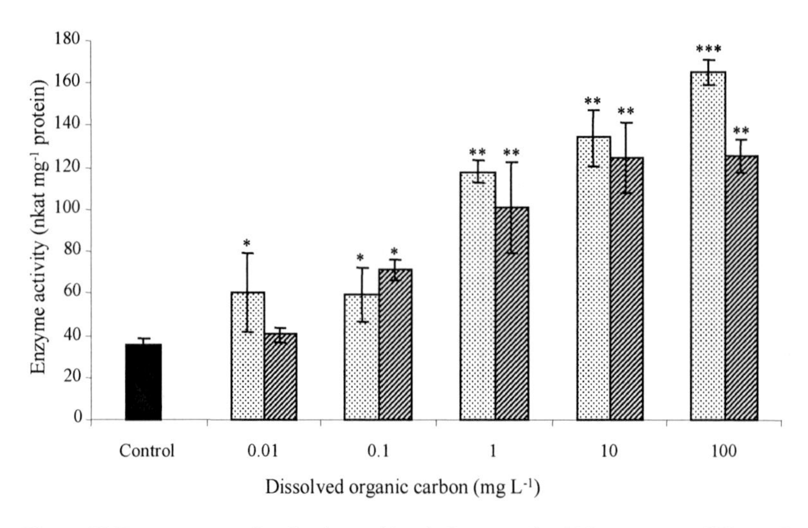

Figure 33. Dose-response of guaiacol peroxidase in *L. minor* after 24 h exposure to different dissolved organic carbon concentrations from two plant leaf-litter extracts. Values are means (n = 4) ± standard deviation, * $p < 0.05$, ** $p < 0.01$, *** $p < 0.001$ (Tukeys HSD test), *P. australis* extract, *Q. robur* extract

3.2.1.6 Enzymatic activity response of catalase in the floating aquatic macrophyte *L. minor*.

In both *Q. robur* and *P. australis* extracts, lower (0.01 and 0.1 mg L^{-1}) DOC levels did not significantly induce the enzymatic activity of catalase (CAT) in *L. minor* compared with control (Fig. 34). However, DOC concentrations ranging from 1 to 100 mg L^{-1} in both extracts led to significant elevation of CAT activity.

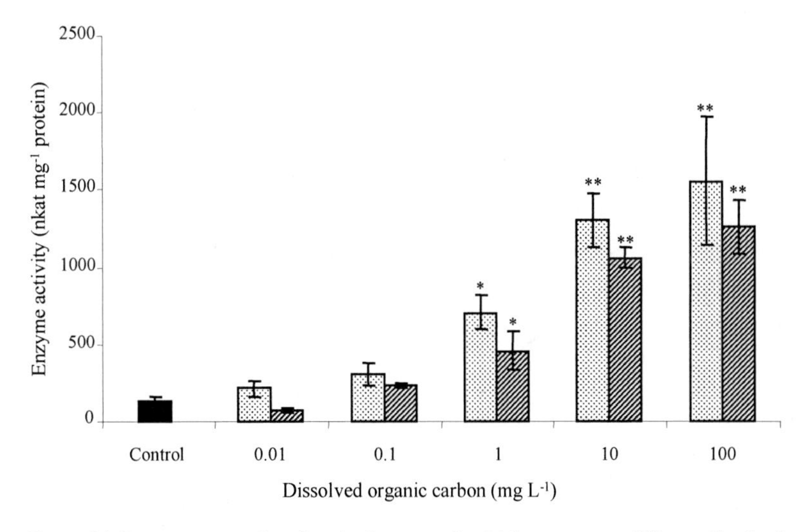

Figure 34. Dose-response of catalase in *L. minor* after 24 h exposure to different dissolved organic carbon concentrations from two plant leaf-litter extracts. Values are means (n = 4) ± standard deviation, * $p < 0.05$, ** $p < 0.01$, *** $p < 0.001$ (Tukeys HSD test), *P. australis* extract, *Q. robur* extract

3.2.1.7 Enzymatic activity response of microsomal glutathione S-transferase (mGST), assayed with the conjugating substrate 1-chloro-2,4-dinitrobenzene (CDNB), in *L. minor*

When exposed to *P. australis* extract, the activity of the insoluble form of glutathione S-transferase (microsomal mGST) in *L. minor* was significantly increased at DOC concentrations of 1, 10 and 100 mg L^{-1} compared with control (Fig. 35). Exposure of *L. minor* to *Q. robur* extract induced mGST activity at all tested DOC concentrations except 0.01 mg L^{-1}. The increase in mGST activity in *Q. robur* extract was more than or up to two orders of magnitude higher than control value at 10 and 100 mg L^{-1} DOC respectively.

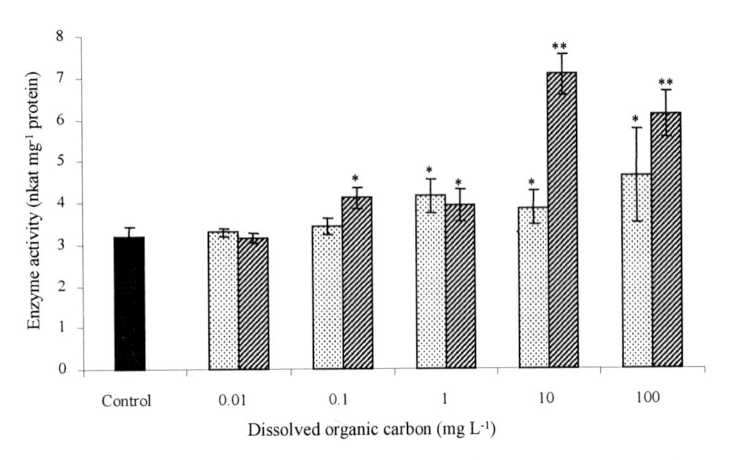

Figure 35. Dose-response of microsomal glutathione S-transferase, measured with CDNB as substrate, in *L. minor* after 24 h exposure to different DOC concentrations from two plant extracts. Values are means (n = 4) ± standard deviation, * $p < 0.05$, ** $p < 0.01$ (Tukeys HSD test), *P. australis* extract, *Q. robur* extract.

3.2.1.8 Enzymatic activity response of cytosolic glutathione S-transferase (cGST), assayed with the conjugating substrate 1-chloro-2,4-dinitrobenzene (CDNB), in *L. minor*.

The activity of the soluble form of glutathione S-transferase (cGST), assayed with CDNB, exhibited relatively lower sensitivity to *P. australis* extract (Fig. 36); whereby significant

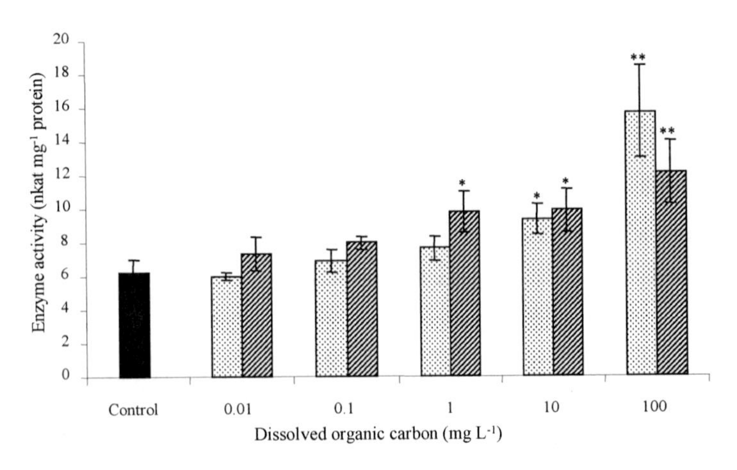

Figure 36. Dose-response of cytosolic glutathione S-transferase, measured with1-chloro-2,4-dinitrobenzene (CDNB) as substrate, in *L. minor* after 24 h exposure to different DOC concentrations from two plant leaf-litter extracts. Values are means (n = 4) ± standard deviation, * $p < 0.05$, ** $p < 0.01$ (Tukeys HSD test), *P. australis* extract, *Q. robur* extracts.

elevation in enzymatic activity was manifested only at higher (10 and 100 mg L^{-1}) DOC levels. In *Q. robur* extract, cGST showed a significant increase in activity at 1, 10 and 100 mg DOC L^{-1}.

3.2.1.9 Enzymatic activity response of cytosolic glutathione S-transferase (cGST), assayed with the conjugating substrate 4-hydroxynonenal (4-HNE), in *L. minor*

Using 4-hydroxynonenal as substrate, soluble cGST in *L. minor* responded similarly as in the case where CDNB was used as substrate. The activity of cGST-4HNE was induced significantly at 10 and 100 mg DOC L^{-1} in *L. minor* exposed to *P. australis* extract (Fig. 37). Significant elevation in enzymatic activity was observed at 1, 10 and 100 mg L^{-1} DOC in *L. minor* treated with *Q. robur* extract.

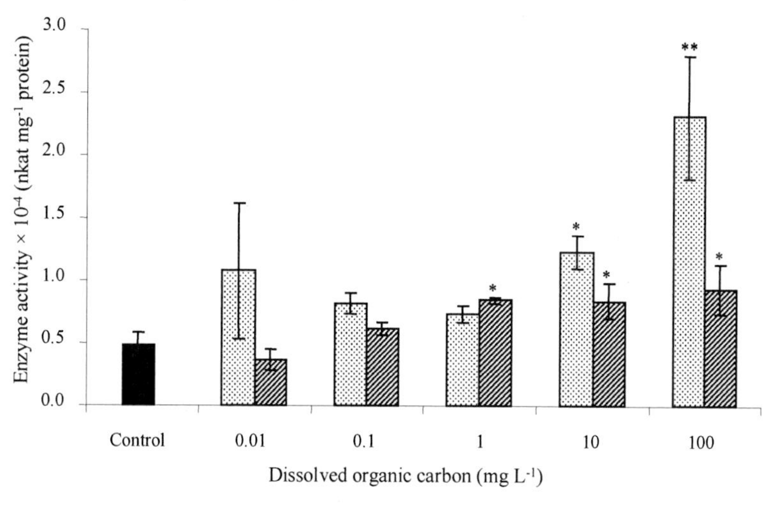

Figure 37. Dose-response of cytosolic glutathione S-transferase, measured with 4-hydroxynonenal as substrate, in *L. minor* after 24 h exposure to different dissolved organic carbon concentrations from two plant leaf-litter extracts. Values are means ($n = 4$) ± standard deviation, * $p < 0.05$, ** $p < 0.01$ (Tukeys HSD test), *P. australis* extract, *Q. robur* extract.

3.2.1.10 Enzymatic activity response of glutathione peroxidase (GPx) in the floating aquatic macrophyte *L. minor*

Glutathione peroxidase in *L. minor* manifested high sensitivity to both extracts. All tested DOC concentrations (except 0.01 mg L^{-1}) elicited induced activity of the GPx enzyme in both extracts with the highest activity observed at the highest DOC level (Fig. 38)

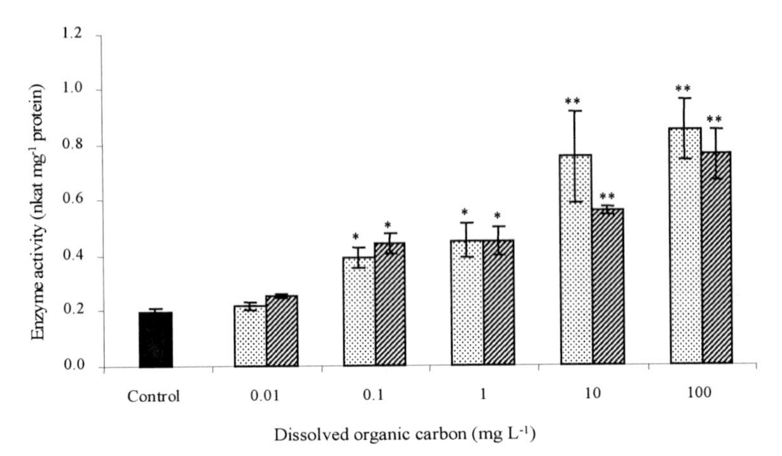

Figure 38. Dose-response of glutathione peroxidase in *L. minor* after 24 h exposure to different dissolved organic carbon concentrations from two plant leaf-litter extracts. Values are means (n = 4) ± standard deviation, * $p < 0.05$, ** $p < 0.01$ (Tukeys HSD test), *P. australis* extract, *Q. robur* extract.

3.2.1.11 Enzymatic response of glutathione reductase in the aquatic macrophyte *L. minor*

Contrary to GPx, glutathione reductase (GR) showed lower sensitivity to both leaf extracts. No significant change in activity of GR occurred at DOC concentrations ranging from 0.01 to 1 mg L^{-1} compared with control values in *L. minor* exposed to both *P australis* and *Q. robur* extracts (Fig. 39).

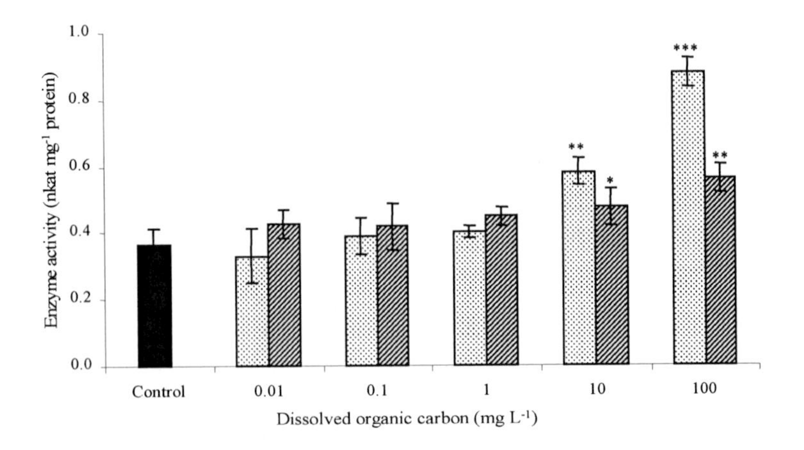

Figure 39. Dose-response of glutathione reductase in *L. minor* after 24 h exposure to different DOC concentrations from two plant leaf-litter extracts. Values are means (n = 4) ± standard deviation, * $p < 0.05$, ** $p < 0.01$ (Tukeys HSD test), *P. australis* extract, *Q. robur* extract.

Significant elevation of GR activity was observed at 10 and 100 mg L^{-1} in both extracts. In *P. australis* extract, GR activity in *L. minor* was by a factor of two higher than that of control (untreated) plants at the highest (100 mg L^{-1}) DOC concentration.

3.2.1.12 Lipid peroxidation in the floating aquatic macrophyte *L. minor*

Lipid peroxidation was evaluated in terms of the lipid peroxidation product malondialdehyde (MDA). The concentration of MDA in treated *L. minor* plants did not change significantly compared with control in all tested DOC concentrations of *Q. robur* extract (Fig. 40). In *L. minor* exposed to *P. australis* extract, MDA concentration significantly increased at the highest (100 mg L^{-1}) DOC concentration. Lower *P. australis* DOC levels did not lead to any significant change in MDA concentration.

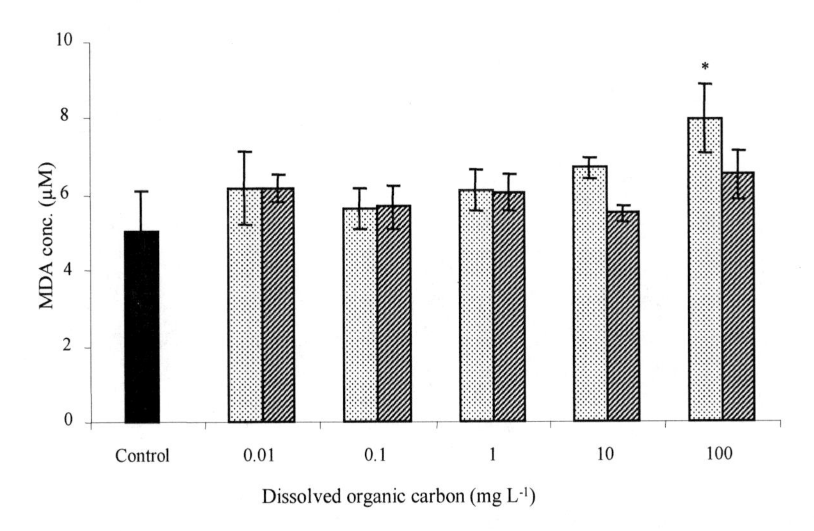

Figure 40. Dose-response of lipid peroxidation in *L. minor* after 24 h exposure to different dissolved organic carbon concentrations from two plant leaf-litter extracts. Values are means (n = 4) ± standard deviation, * $p < 0.05$ (Tukeys HSD test), *P. australis* extract, *Q. robur* extracts.

3.2.2 Experiment 5: Acclimation experiments with *L. minor* exposed to *P. australis* and *Q. robur* extracts

3.2.2.1 Enzymatic response of guaiacol peroxidase (POD) on time-dependent exposure of *L. minor* to *P. australis* and *Q. robur* extracts

The profile of guaiacol peroxidase (POD) in *L. minor* treated with 10 mg L^{-1} DOC from *P. australis* and *Q. robur* leaf litter extracts is shown in Figure 41. The results show that the activity of POD enzyme was not significantly modified within 12 h of exposure to *P. australis* extracts. Significant elevation in POD activity due to *P. australis* extract occurred after 24, 48 and 168 h. On the other hand, *Q. robur* extract induced a significant time-dependent increase of POD activity in *L. minor* after 8 h.

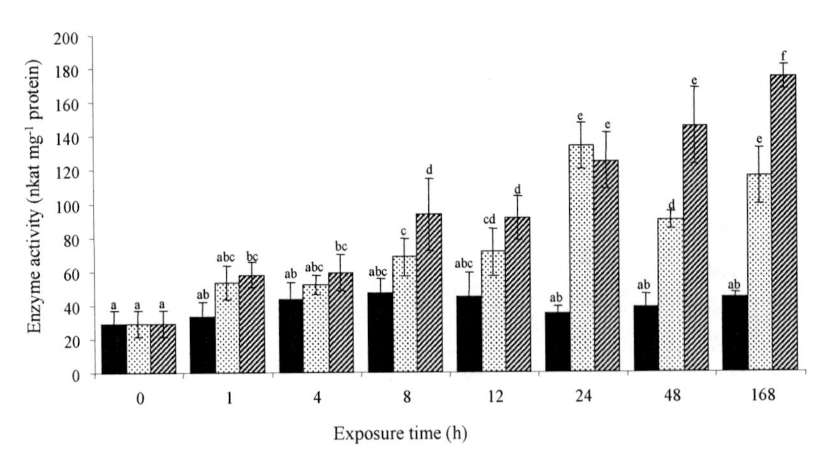

Figure 41. Time-response of guaiacol peroxidase activity in *L. minor* after exposure to 10 mg L^{-1} DOC from *P. australis* and *Q. robur* extracts and Control. Values with the same letter are not significantly different (Tukeys HSD test), n = 4.

3.2.2.2 Enzymatic response of catalase on time-dependent exposure of *L. minor* to *P. australis* and *Q. robur* extracts

Both *P. australis* and *Q. robur* leaf litter extracts significantly induced the activity of catalase (CAT) in *L. minor* after 8 h. A further increase in CAT activity occurred at 12 h and reached a maximum at 24 h in both extracts. At 48 and 168 h, *L. minor* exposed to *P. australis* extract manifested a sharp decrease in CAT activity compared with the level noted at 24 h. In *Q. robur* extract, CAT activity reached a plateau at 48 h and started to decrease after 168 h compared with activity level observed at 24 h. Although CAT activity started to decrease after 48 and 168 h, the activity remained elevated in comparison with control levels in both extracts (Fig. 42).

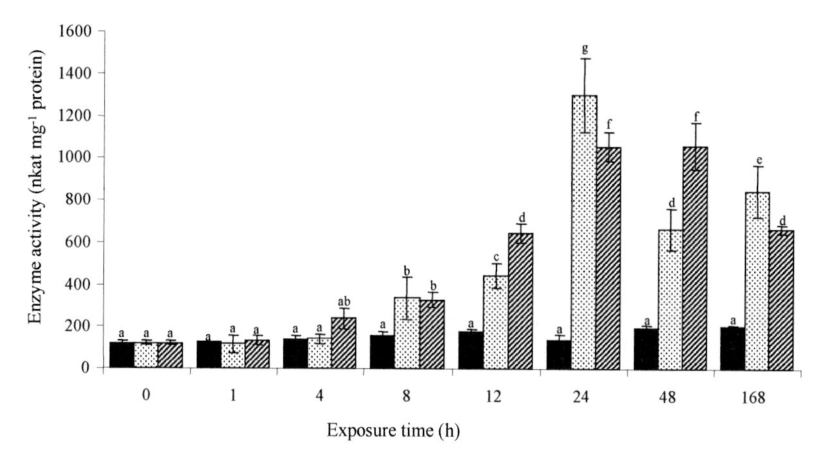

Figure 42. Time-response of catalase activity in *L. minor* after exposure to 10 mg L^{-1} DOC from *P. australis* and *Q. robur* extracts and control. Values with the same letter are not significantly different (Tukeys HSD test), n = 4.

3.2.2.3 Enzymatic response of microsomal glutathione S-transferase (mGST), measured with 1-chloro-2,4-dinitrobenzene (CDNB), on time-dependent exposure of *L. minor* to *P. australis* and *Q. robur* extracts

P. australis extract elicited a delayed but time-dependent increase in activity of the membrane-bound (microsomal) glutathione S-transferase (mGST) enzyme starting at 12 h and continued till the end of the experiment (168 h) (Fig. 43). *L. minor* exposed to *Q. robur* extract showed a significant increase in mGST activity after 8 h, reached a maximum at 24 h and decreased slightly but still remained higher than control values at 48 and 168 h.

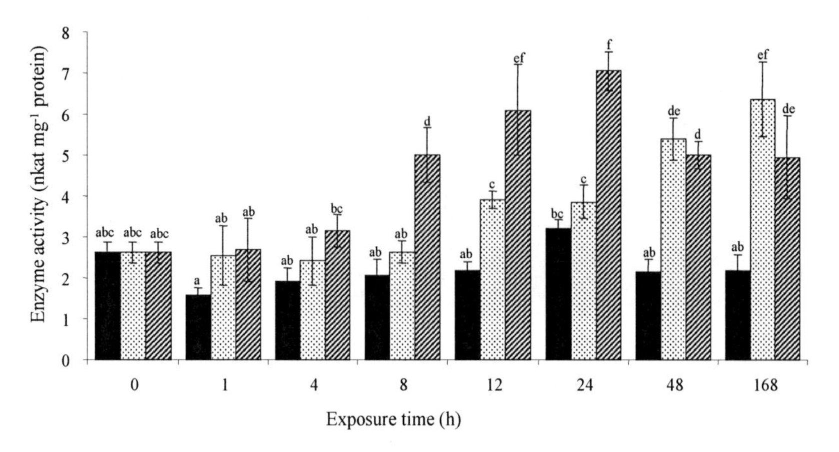

Figure 43. Time-response of microsomal glutathione S-transferase (mGST), measured with CDNB, in *L. minor* after exposure to 10 mg L^{-1} DOC from *P. australis* and *Q. robur* extracts and control. Values with the same letter are not significantly different (Tukeys HSD test), n = 4.

3.2.2.4 Enzymatic response of cytosolic glutathione S-transferase (cGST), measured with CDNB, on time-dependent exposure of *L. minor* to *P. australis* and *Q. robur* extracts

Soluble (cytosolic) glutathione S-transferase assayed with CDNB (cGST-CDNB) exhibited a different pattern of response to the two extracts, attaining two maxima in *Q. robur* extracts at 8 and 48 h respectively. In *P. australis* extract, a significant increase in cGST-CDNB activity, in comparison with control values, was not observed until after 24 h exposure and remained elevated till the end of the experiment (168 h) (Fig. 44). In *Q. robur* extract, cGST-CDNB activity was significantly induced after 4 h exposure, reached an initial maximum at 8 h, then declined slightly but remained higher than control values at 12 and 24 h. The activity of cGST-CDNB increased again and reached maximum at 48 h in *Q. robur* extracts.

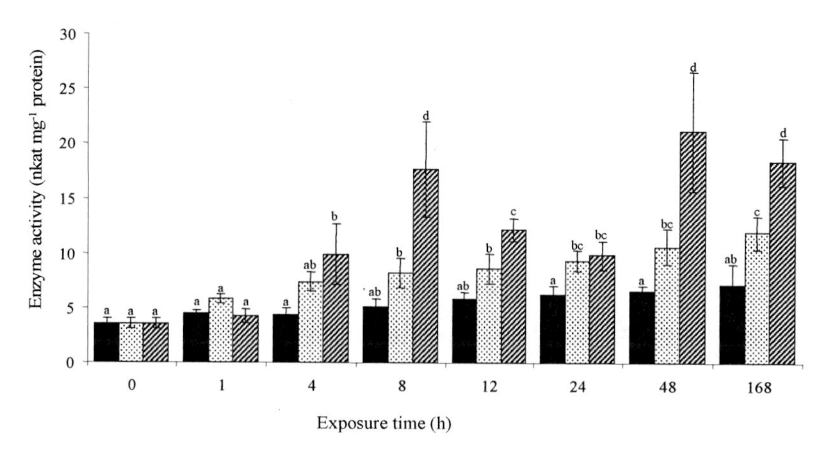

Figure 44. Time-response of cytosolic glutathione S-transferase (cGST), measured with CDNB, in *L. minor* after exposure to 10 mg L^{-1} DOC from *P. australis* and *Q. robur* extracts and control. Values with the same letter are not significantly different (Tukeys HSD test), n = 4.

3.2.2.4 Enzymatic response of cytosolic glutathione S-transferase (cGST), measured with 4-hydroxynonenal, on time-dependent exposure of *L. minor* to *P. australis* and *Q. robur* extracts

When measured with 4-hydroxynonenal, soluble (cytosolic) glutathione S-transferase (cGST-4HNE) was significantly induced after 12 h exposure of *L. minor* to both *P. australis* and *Q. robur* extracts (Fig. 45).

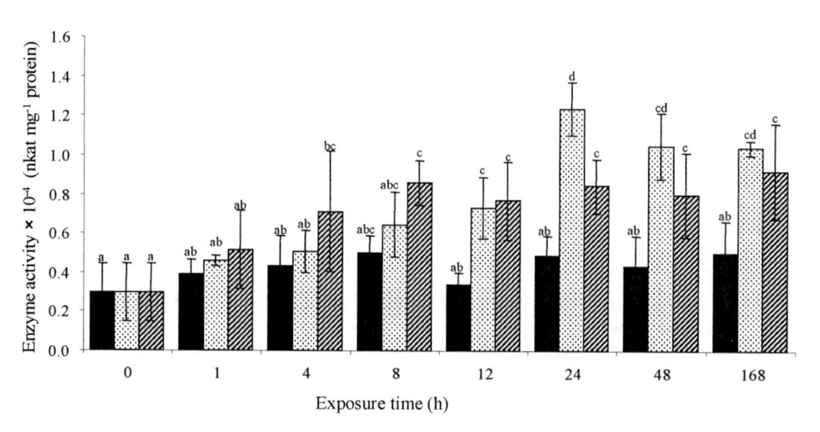

Figure 45. Time-response of cytosolic glutathione S-transferase (cGST), measured with 4-HNE, in *L. minor* after exposure to 10 mg L^{-1} DOC from *P. australis* and *Q. robur* extracts and control. Values with the same letter are not significantly different (Tukeys HSD test), n = 4.

A further increase of enzyme activity in *P. australis* extract occurred at 24 h and then levelled off till the end of the experiment. Compared with the activity noted at 8 h, no further increase occurred up to the end of the experiment in *L. minor* exposed to *Q. robur* extract.

3.2.2.5 Enzymatic response of glutathione peroxidase on time-dependent exposure of *L. minor* to *P. australis* and *Q. robur* extracts

As shown in Figure 46, exposure of *L. minor* to 10 mg L^{-1} DOC leaf litter extracts provoked no significant changes in glutathione peroxidase (GPx) activity within 8 h (*Q. robur* extract) and 12 h (*P. australis* extract). At 24 h, *P. australis* extract caused significant increase in GPx activity, followed by a decrease in activity at 48 and 168 h relative to the activity noted at 24 h but did not return to control levels. For *L. minor* exposed to *Q. robur* extract, a significant elevation in GPx activity occurred at 12 h with a further increase at 48 h which was sustained till the end of the experiment (168 h).

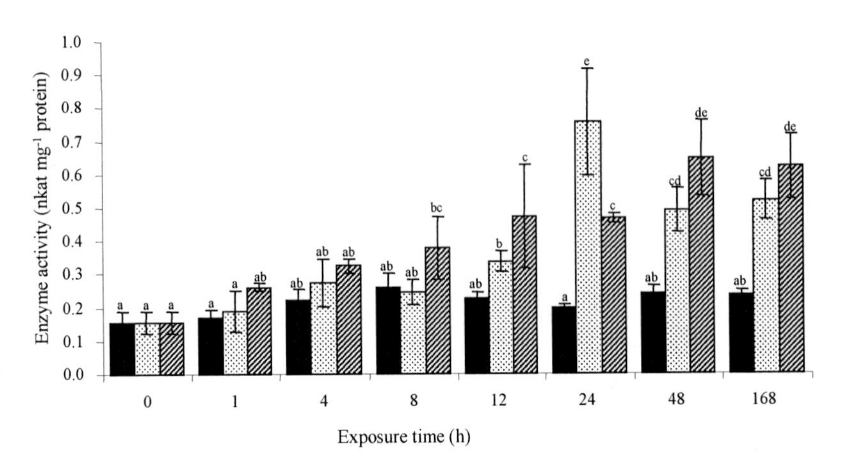

Figure 46. Time-response of glutathione peroxidase activity in *L. minor* after exposure to 10 mg L^{-1} DOC from [dotted] *P. australis* and [hatched] *Q. robur* extracts and [black] Control. Values with the same letter are not significantly different (Tukeys HSD test), n = 4.

3.2.2.5 Enzymatic response of glutathione reductase on time-dependent exposure of *L. minor* to *P. australis* and *Q. robur* extracts

Glutathione reductase (GR) displayed the most sensitive and variable responses to *P. australis* extract exposure with a significantly induced activity after 1 h and sustained throughout the experiment (Fig. 47). At 24 h in *P. australis* extract, GR activity declined relative to the preceding activity noted at 12 h, and then increased to the maximum level at 48 h. In *Q. robur* extract, GR activity was significantly induced at 4 h and remained sustained throughout the experiment.

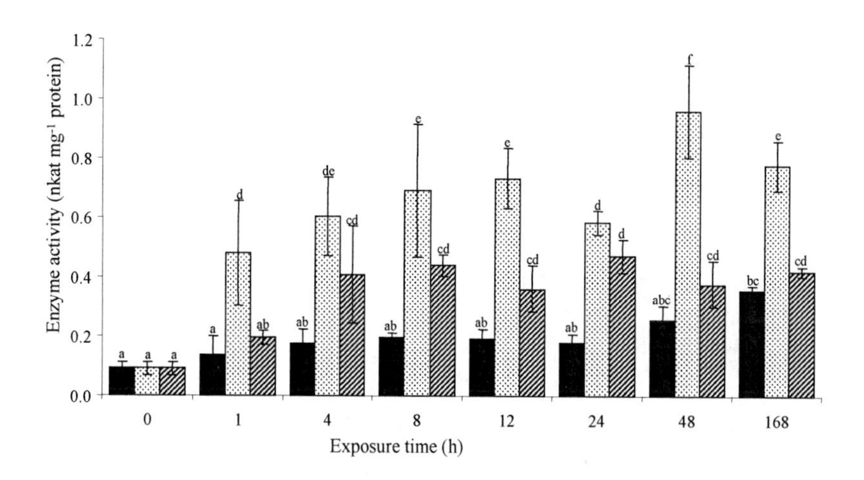

Figure 47. Time-response of glutathione reductase activity in *L. minor* after exposure to 10 mg L^{-1} DOC from *P. australis* and *Q. robur* extracts and Control. Values with the same letter are not significantly different (Tukeys HSD test), n = 4.

3.2.3 Experiment 6: *L. minor* bioassay experiments

3.2.3.1 Chlorophyll pigments in *L. minor* exposed to *P. australis* and *Q. robur* extracts

After 3 days (72 h), the total chlorophyll content was not significantly reduced except at the highest tested DOC concentration (100 mg L^{-1}) in *Q. robur* extract (Fig. 48a). After 5 days (120 h) of exposure, significant reduction in total chlorophyll concentration was observed at 10 and 100 mg DOC L^{-1} *Q. robur* extracts. DOC levels below 10 mg L^{-1} in *Q. robur* extracts did not result in any significant reduction in pigment contents within the investigated period (7 days). Within 5 days of exposure to *P. australis* extracts, significant decrease of total chlorophyll content occurred at 10 and 100 mg DOC L^{-1} (Fig. 48b). At day 7, chlorophyll content in *L. minor* exposed to *P. australis* extract decreased significantly at 1 mg L^{-1} DOC.

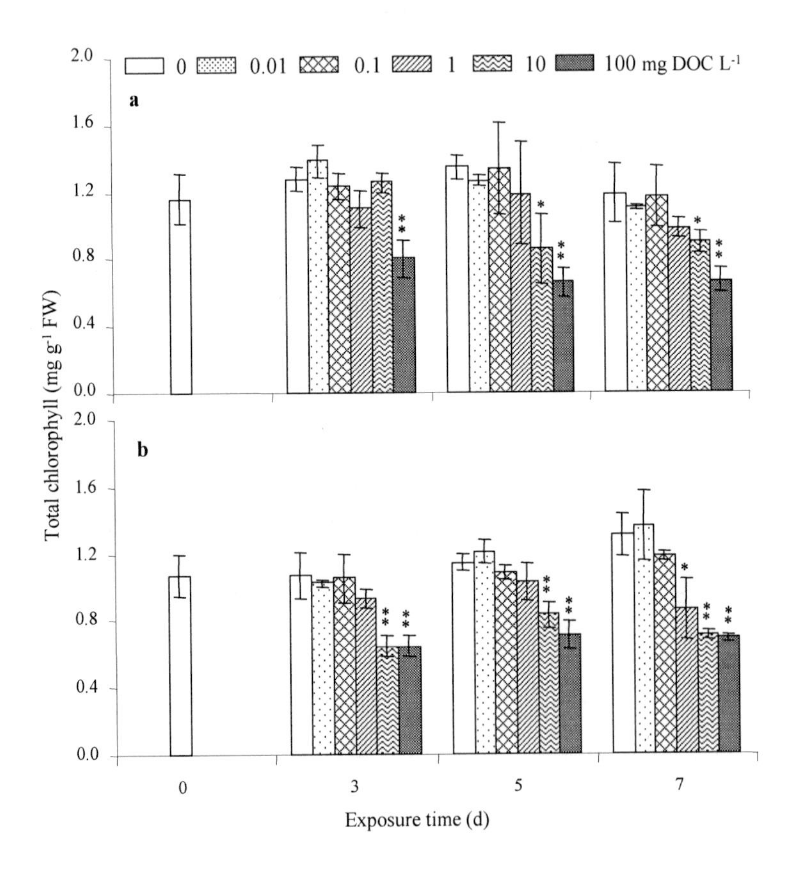

Figure 48. Total chlorophyll content (mg g^{-1} FW) in *L. minor* exposed to various DOC concentrations (mg L^{-1}) of (a) *Q. robur* and (b) *P. australis* extracts. Bars represent mean (n = 4) ± SD, * $p<0.05$, ** $p<0.01$ (Newman-Keuls test).

Chlorophyll *a/b* ratio significantly declined at the highest tested DOC level (100 mg L^{-1}) in *L. minor* exposed to *Q. robur* extracts throughout the experiment (Fig. 49a). In *P. australis* extracts, significant reduction in chlorophyll *a/b* ratio occurred at 10 and 100 mg L^{-1} DOC throughout the experiment (Fig. 49b).

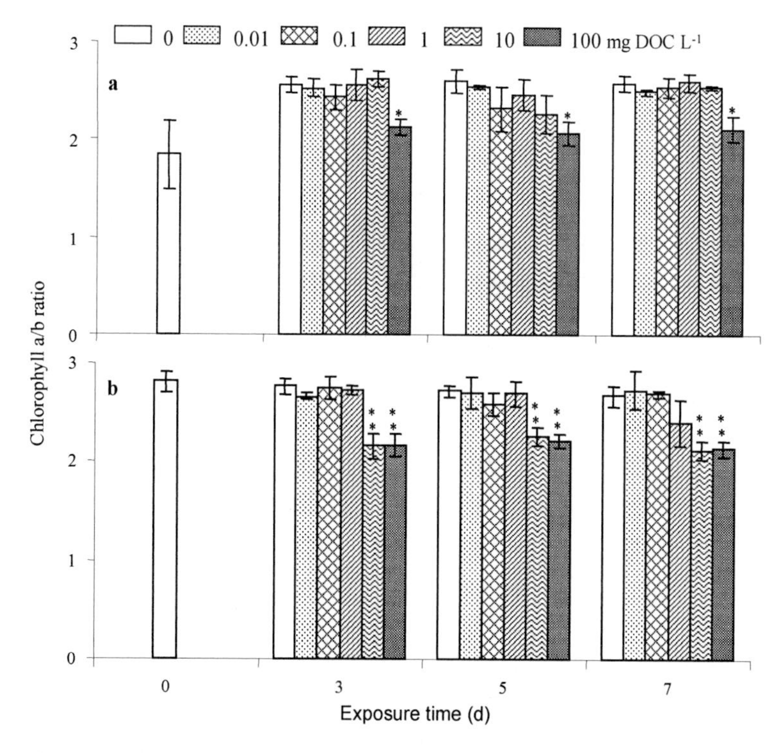

Figure 49. Chlorophyll a/b ratio in *L. minor* exposed to various DOC concentrations (mg L^{-1}) of (a) *Q. robur* and (b) *P. australis* extracts. Bars represent mean (n = 4) ± SD, * $p<0.05$, ** $p<0.01$ (Newman-Keuls test).

3.2.3.2 Photosynthesis in *L. minor* exposed to *P. australis* and *Q. robur* extracts

Photosynthetic reduction in *L. minor* exposed to *Q. robur* extract was observed at the highest DOC concentration (100 mg L^{-1}) within the first 3 days (Fig 50a). Further on (day 5) in *Q. robur* extract, photosynthetic inhibition occurred at 1, 10 and 100 mg DOC L^{-1}. At the end of the experiment (7 days), *Q. robur*-derived DOC level as low as 0.1 mg L^{-1} evoked significant reduction in the photosynthetic capacity of *L. minor*. On the other hand, *Lemna* exposed to *P. australis* extracts exhibited different photosynthetic responses at the early and later stages of exposure. On day 3, significant photosynthetic inhibition occurred at 10 and 100 mg L^{-1} DOC *P. australis* extract (Fig. 50b). On the fifth and seventh days of exposure, *P. australis* extract inhibited photosynthesis at 1, 10 and 100 mg L^{-1} DOC.

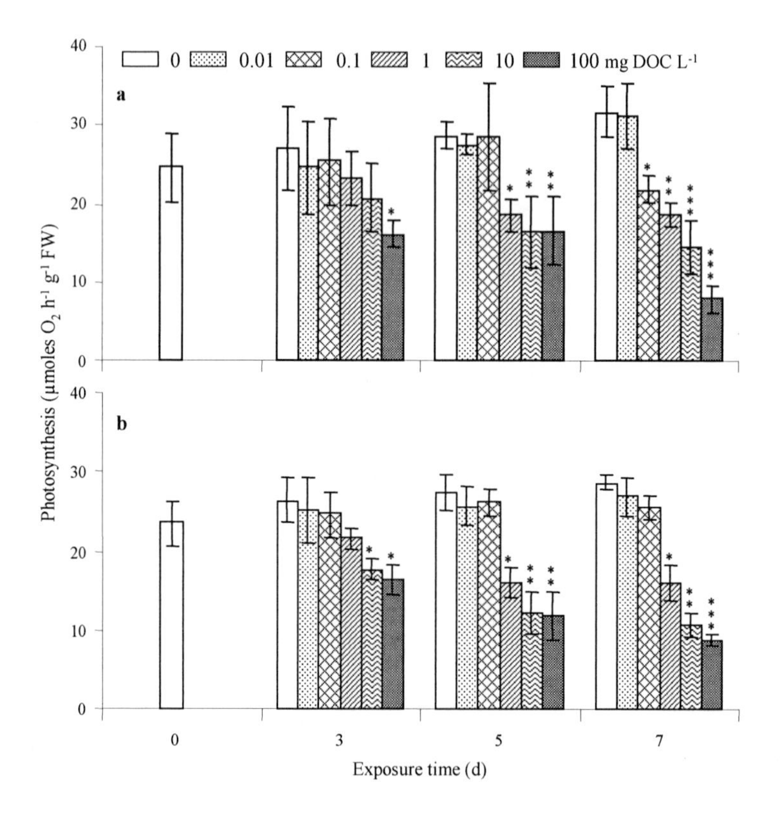

Figure 50. Photosynthetic oxygen release in *L. minor* after exposure to various DOC concentrations (mg L^{-1}) of (a) *Q. robur* and (b) *P. australis* leaf litter extracts. Bars represent mean (n = 4) ± SD, * p<0.05 ** p<0.01 ***p<0.001 (Newman-Keuls test).

3.2.3.3 Number of fronds in *L. minor* exposed to *P. australis* and *Q. robur* extracts

Regarding absolute frond numbers, the inhibitory effect in *Q. robur* extracts was time and concentration dependent. After 3 and up to 5 days of exposure, *Q. robur* extracts led to significant decrease in frond numbers at DOC concentrations of 1, 10 and 100 mg L^{-1} (Fig. 51a). After 7 days of exposure to *Q. robur* extract, a significant decrease was also observed at 0.1 mg L^{-1} DOC.

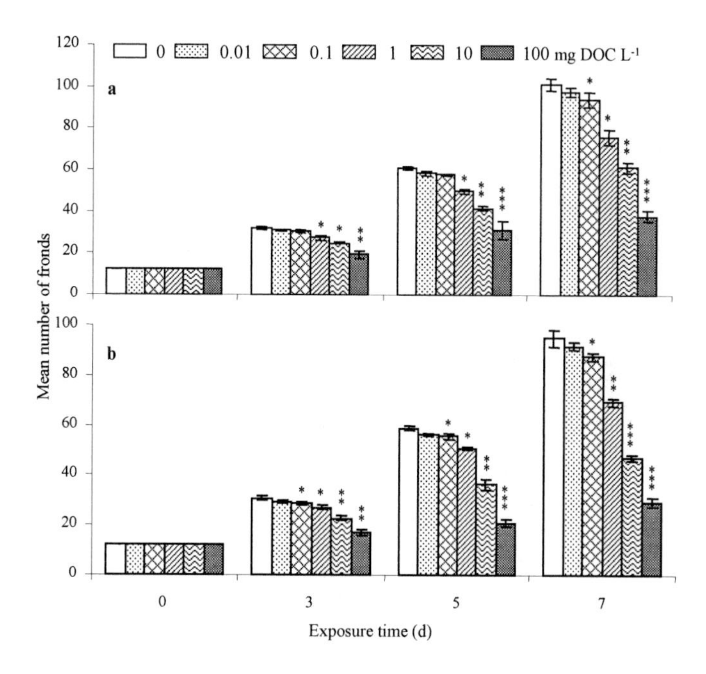

Figure 51. Frond numbers in *L. minor* exposed to various DOC concentrations (mg L^{-1}) of (a) *Q. robur* and (b) *P. australis* extracts. Bars represent mean (n = 4) ± SD, * $p<0.05$, ** $p<0.01$, *** $p<0.001$ (Newman-Keuls test).

In *P. australis* extract (Fig. 51b), changes in frond numbers as a function of DOC concentration and exposure time was variable, the inhibitory effects being dose but not time-dependent. Already after 3 days of exposure to *P. australis* extract, significant reduction in frond number compared with control value was observed at all but the lowest (0.01 mg L^{-1}) DOC concentration; this remained unchanged up to the end of the experiment (7 days).

3.2.3.4 Growth rates in *L. minor* exposed to *P. australis* and *Q. robur* extracts

Lemna exhibited different growth inhibitory responses to the two leaf extracts depending on the endpoint considered. Based on dry weight (DW), *Q. robur* extract significantly inhibited the growth rate of *Lemna* after 7 days of exposure at all tested DOC concentrations (except 0.01 mg L^{-1}), with the strongest effect observed at 100 mg L^{-1} which was equal to 53.9% reduction over control levels (0 mg L^{-1} DOC) (Fig. 52a).

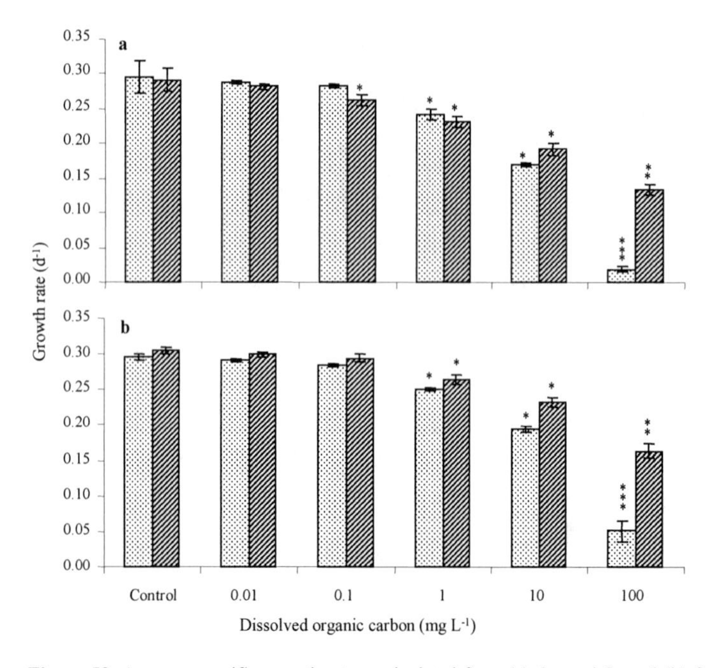

Figure 52. Average specific growth rates, calculated from (a) dry weight and (b) frond number, of *L. minor* after 7 days exposure to various DOC concentrations. Bars represent mean (n = 4) ± SD, * p<0.05 ** p<0.01 ***p<0.001 (Newman-Keuls test), *P. australis* extract, *Q. robur* extracts.

For *P. australis* extract, significant reduction in DW-specific growth rate started at 1 mg L^{-1} DOC. The strongest inhibition of DW-based growth rate in *L. minor* exposed to *P. australis* extract occurred at 100 mg DOC L^{-1} and corresponded to 61.3%. Considering frond number (FN)-specific growth rate (Fig. 52b), both extracts caused a significant reduction in growth rate at 1, 10 and 100 mg L^{-1} DOC. Growth rate inhibition was higher in *P. australis* than in *Q. robur* extract at high DOC concentrations regardless of the endpoint considered.

3.2.3.5 Toxicity dose-response analyses in *L. minor* exposed to *P. australis* and *Q. robur* extracts

The 7 days *Lemna* growth inhibition data for *Q. robur* and *P. australis* based on dry weight and frond number were used to perform standard toxicity analyses and the results are presented in Table 8. A four parameter sigmoidal concentration-response curve-fitting model was used to estimate median growth effect (EC_{50}) concentrations and confidence intervals for the two extracts. In *L. minor* exposed to *Q. robur* extract, the EC_{50} was determined to be 63.8

(DW) and 148 (FN) mg L^{-1}. In *P. australis* extracts, the EC_{50}s were 28.4 (DW) and 49.4 (FN) mg L^{-1}. At lower DOC concentrations, *L. minor* showed similar sensitivity to *Q. robur* and *P. australis* extracts in terms of FN-specific lowest observed effect concentration (LOEC) (1 mg L^{-1}) and no observed effect concentration (NOEC) (0.1 mg L^{-1}). Regarding DW-specific growth rate, *L. minor* manifested different responses to the two extracts at low DOC concentrations. In *Q. robur* extract, DW-specific LOEC and NOEC values were 0.01 and 0.1 mg L^{-1} DOC respectively, while in *P. australis* extracts LOEC and NOEC values of 0.1 and 1 mg L^{-1} respectively, were determined.

Table 8. 168 h median, lowest and no observed effect concentrations (EC_{50}, LOEC and NOEC) in mg L^{-1} DOC of tested *Q. robur* and *P. australis* leaf litter extracts determined from different growth endpoints in *L. minor* exposures.

	EC_{50}	95% CI	R^2	LOEC	NOEC
Oak					
Dry weight	63.8	50.5 - 80.7	0.985	0.1	0.01
Frond number	148.0	114.2 - 191.8	0.981	1.0	0.10
Reed					
Dry weight	28.4	23.0 - 35.0	0.983	1.0	0.10
Frond number	49.4	42.6 - 57.2	0.992	1.0	0.10

4. Discussion

4.1 Discussion: Experiments with *Ceratophyllum demersum*

4.1.1 Physio-biochemical response experiments

4.1.1.1 Experiment 1 (Short-term 24 h dose-dependent): Response of the photosynthetic and antioxidative system of *C. demersum* to leaf litter-derived DOC

Photosynthesis is an important physiological process often used to test the sensitivity of aquatic plants to changes in their surrounding (Sersen et al., 1998; Menone and Pflugmacher, 2005). In the present study, a significant concentration-dependent reduction in the rate of photosynthesis of *C. demersum* after a 24 h exposure to both *Q. robur* and *P. australis* extracts was observed. Similar effects have been demonstrated in earlier studies using humic substances isolated from soils (Pflugmacher et al., 1999). The mechanism of action for this observation is not clear, as there was no detectable effect on chlorophyll pigments. However, aromatic structures like phenols and other aromatic hydrocarbons are known to have toxic effects on organisms. For example the oxidation of quinone leading to the formation of semiquinone radicals is known to interfere with the electron transfer chain in photosystem II (PSII) in plant chloroplasts (Oettmeier et al., 1988, Pflugmacher et al., 2006). Using the pure quinone, 1-aminoanthraquinone, photosynthetic oxygen release in *Vesicularia dubyana* was reduced to almost zero (Pflugmacher et al., 2003). It was also shown in a recent study that the aromatic organic compound, 3-chlorobiphenyl, significantly inhibited the photosynthesis of *C. demersum* at a concentration of 5 mg L^{-1} (Menone and Pflugmacher, 2005). The significant ($P<0.01$) reduction in rate of photosynthetic oxygen release due to the oak (*Q. robur*) extract, which is well known for its high tannin and phenolic contents (Salminen et al., 2004), suggests a possible interference with electron transport chain from PSII to PSI (Pflugmacher et al., 2006). The adverse effect caused by *Q. robur* extract was indeed greater than that caused by *P. australis* extract. Additionally, it is known that when plants are exposed to stress conditions, there is an increase in ROS (Pflugmacher, 2004) and the antioxidant capacity of the plant may be overwhelmed. Organelles such as the peroxisomes and chloroplast (site of photosynthesis), where ROS are also produced under 'normal' conditions, are especially at risk (Alscher, 2002). Overproduction of ROS in chloroplasts of plants under drought-stress has been reported (Price et al., 1989), and a similar mechanism may occur when plants are exposed to certain concentrations of leaf extracts. The induced activity of antioxidative enzymes observed in the present study suggests an increase in ROS, and the susceptibility of the chloroplasts to high ROS could be an additional factor contributing to the corresponding reduction in the rate of photosynthesis.

ROS such as $O_2^{\cdot -}$, $OH^{\cdot}$, H_2O_2 are often internally formed as products of normal plant metabolism (Foyer et al., 1994). Under normal habitat conditions, the rate at which ROS are formed by plants is in dynamic equilibrium with the rate at which they are further utilized or broken down. Leaf extracts might be able to promote oxidative stress by inducing the production of more ROS. To combat excess ROS, plants have developed an antioxidative defense system. Parts of this system are the guaiacol peroxidases (POD), glutathione S-transferases (GST) and glutathione reductase (GR).

In the present study, there was a significant ($P < 0.01$) increase in POD activity in the aquatic macrophyte *C. demersum* after exposure to both *Q. robur* and *P. australis* extracts for 24 h. Exposure of the macrophyte to the leaf extracts may have resulted in an increase in the production of H_2O_2 beyond the normal detoxifying capacity of the plant. In order to deal with this high level of H_2O_2 and to prevent or minimize damage to the plant cells, the plant POD enzymes increased their activity many folds thereby catalyzing the conversion of the highly reactive H_2O_2 into harmless water molecules. Previous studies with other plant species such as *Hydrilla verticillata* also demonstrated a significant increase in POD activity after exposure to anthracene in concentrations above 0.01 mg L^{-1} (Byl et al., 1994). Using humic substances from different soil and water sources, Pflugmacher et al. (1999) also showed an increase in POD activity on *C. demersum* after 7 days exposure in a concentration of 0.5 mg L^{-1}. The very high increase (up to 47 folds) in POD activity observed in the present study was after 24 h exposure of *C. demersum* to crude leaf extracts. DOC concentrations as low as 0.1 mg L^{-1} provoked significant ($P < 0.01$) increases in POD activity. Aromatic structures and plant growth hormones like gibberellic acid are known to cause an elevation of POD activity (Oberg et al., 1990, Kwak et al., 1996). Response of POD activity upon exposure to leaf extracts from *P. australis* and *Q. robur* suggests that these chemicals may contain similar structures. *Q. robur*, for example, is well known for its high tannin and phenolic contents (Salminen et al., 2004). The antioxidant activity of phenols is mainly due to their redox properties, which allow them to act as reducing agents, hydrogen donors and singlet oxygen quenchers (Rice- Evans et al., 1995).

Soluble and microsomal glutathione S-transferase (cGST and mGST) activity was also found to be significantly elevated upon exposure to extracts from *Q. robur* and *P. australis*. These enzymes are invariably involved in cellular biotransformation when plants are faced with xenobiotic or other toxic substances. One of the important mode of action is the conjugation of toxicants or xenobiotics to the sulphurhydryl (–SH) group of glutathione (GSH) which enhances their water solubility and therefore their excretion from cells through cell membranes, incorporation into cell wall or compartmentation into vacuoles or vessicles. Increase in GST activity is thus considered as a stress signal. Similarly, Menone and Pflugmacher (2005) demonstrated that soluble GST activity in *C. demersum* was significantly elevated when treated with lower concentrations (up to 0.5 mg L^{-1}) of the organic PCB compound 3-chlorobiphenyl. Imposition of oxidative stress often leads to the conversion of the existing pool of reduced glutathione (GSH) to the oxidized form, glutathione disulphide (GSSG), thereby stimulating glutathione biosynthesis (May and Leaver, 1993). After GSH has been oxidized to GSSG, the recycling of GSSG to GSH is accomplished mainly by the enzyme glutathione reductase (GR). An elevated GR activity detected in the present study was a further indication of an oxidative stress condition in the exposed *C. demersum*. GSH is used by glutathione peroxidise (GPx) to detoxify hydrogen peroxide and in the process is converted to GSSG, which is then recycled to GSH by GR. The regenerated GSH is then available for the detoxification of more hydrogen peroxide. An increase in GSH biosynthetic capacity has been shown to enhance resistance to oxidative stress (Zhu et al., 1999).

ROS are extremely reactive and unstable chemical species, which react with proteins, nucleic acids, carbohydrates, and lipids in cells. The latter often result in lipid peroxidation. Malondialdehyde (MDA) is one of the end products of lipid peroxidation. The findings in the present study did not reveal any significant ($P < 0.05$) increase in MDA levels in exposed *C. demersum* in all concentrations tested. This is presumably due to activation of defense and detoxification mechanisms, as evidenced by high GST, GR and POD activities. These antioxidant enzymes can enhance the removal of ROS and prevent damage to cellular constituents and initiation of lipid peroxidation (Yang et al., 2003). Furthermore, GST conjugates MDA and 4-hydroxynonenal (4-HNE) using GSH as substrate in the biotransformation process.

In conclusion, these results indicate that leaf extracts have the capacity to impose physiological stress and provoke acute antioxidative responses in the aquatic macrophyte *C. demersum*. The effects observed within 24 h are quite strong. This means that aquatic macrophytes could serve as early warning systems in water quality assessment procedures.

Furthermore, it was shown that even relatively low DOC concentrations of leaf extracts from *P. australis* and *Q. robur* can induce oxidative stress as well as slow down the rate of photosynthesis in the coontail, *C. demersum*. The effects tend to level off or in some cases start to decrease at high DOC concentrations indicating a threshold on the antioxidative ability of the macrophyte. In this study, it seems evident that the antioxidative system in *C. demersum* was sufficiently activated to prevent or minimize this damage since no significant effect on lipid peroxidation was detected in treated plants. These findings thus suggest that aqueous leaf extracts may be an important environmental factor affecting aquatic biota and contributing to the overall ecological dynamics in shallow lakes and rivers.

4.1.1.2 Experiment 2: Acclimation (time response) of *Ceratophyllum demersum* to the effects of leaf litter-derived DOC

C. demersum was exposed for different times to 10 mg L^{-1} DOC of *Q. robur* and *P. australis* extracts because the strongest effects were, in most cases, observed at this concentration. The aim was to monitor the photosynthetic efficiency and enzyme kinetics through time in order to assess the acclimatory potential of *C. demersum* to the effects of leaf litter extracts.

Exposure of *C. demersum* to *Q. robur* and *P. australis* leaf extracts results in different physiological responses, one being reversible and the other not. Upon exposure to *P. australis* extract, the assumption of a phenotypic plasticity allowing an acclimation of photosynthesis to DOC toxicity was evoked. In the case of *Q. robur* extract, it seems that plasticity was overwhelmed. The onset of photosynthetic inhibition in *C. demersum* occurred relatively sooner upon exposure to *P. australis* extracts (4 h) than on exposure to *Q. robur* extracts (12 h). Photosynthetic recovery in *C. demersum* started after 48 h exposure to *P. australis* leaf extracts. At the end of the experiment (after 168 h), normal photosynthetic performance was restored in *C. demersum* exposed to *P. australis* extract. On the other hand, photosynthetic O_2 release in *C. demersum* exposed to *Q. robur* extract did not recover within the period considered in this study. Failure of the photosynthetic system in *C. demersum* to recover after a relatively prolonged (one week) exposure to *Q. robur* extract may be attributed to its chemical composition. *Q. robur* is well known for its relative high tannin contents and humic substances with aromatic and quinoide structures (Salminen et al., 2004) that may interfere with the electron transport chain of photosystem II in the plant chloroplast (Oettmeier et al.,

1988; Pfannschmidt, 2003). Although photosynthetic-acclimation response may vary depending on the plant species and/or plant part (Madsen et. al., 1996; Madsen and Brix, 1997), our findings indicate that photosynthetic plasticity is also a function of the stress factor (e.g. DOC) taken into account. Previous studies also demonstrated some photosynthetic plasticity when exposing plants to other stress factors like CO_2 levels (Madsen et. al., 1996), light intensity and water stress (Maxwell et. al., 1994), temperature and inorganic carbon (Olesen and Madsen, 2000). The present study focused on the effect of DOC from leaf litter and associated biofilms often involved in leaves decomposition following breakdown and shredding by invertebrates (Hieber and Gessner, 2002). Photosynthetic recovery in *C. demersum* exposed to *P. australis* extract occurred after 7 days (168 h) suggesting photosynthetic acclimation. As previously observed (Kamara and Pflugmacher, 2007a), the antioxidative defense system was in parallel induced due to stress imposed by exposure to the leaf extracts. The biotransformation enzyme, glutathione S-transferase, was on the one hand significantly induced within one hour of exposure to *Q. robur* extract. On the other hand, the onset of GST response in *C. demersum* upon exposure to *P. australis* extract was variable. The membrane bound form (mGST) was induced within 1 h while soluble GST response, depending on the substrate, started after 4 h (4-HNE substrate) or after 8 h (CDNB substrate) of exposure to *P. australis* extract. However, *P. australis* extract provoked a sooner (after 8 h) elevation in GR activity compared with *Q. robur* extract (after 24 h), both of which effects were not sustained beyond 24 h. GR thus represents the exceptional case in the enzymatic response of *C. demersum* in which *P. australis* extract elicited an earlier response than *Q. robur* extracts. The early response of the other enzymes of the antioxidative system (cGST, POD, GPx) seems to suggest that *C. demersum* was probably more sensitive to *Q. robur* than to *P. australis* extract. If so, this may invoke an earlier *de novo* biosynthesis of GSH (Noctor et. al., 1998) which participates in a variety of ROS-scavenging biochemical reactions as well as being directly involved in the detoxication of certain xenobiotics during stress (Rennenberg, 1982). This might explain why GR, which replenish depleted GSH by reducing glutathione disulphide (GSSG), was elevated later in *Q. robur* extracts, probably after GSH biosynthetic pathway was overwhelmed. GR in *C. demersum* exhibited complete recovery at 168 h in both extracts. For the other enzymes (GST, POD, GPx), it depends on the exposure; *P. australis* extract led to complete recovery while for *Q. robur* extract, the recovery was partial. Overall, antioxidative enzymes in *C. demersum* were more sensitive to *Q. robur* than to *P. australis* extract since they tended to increase sooner (except GR and mGST) and reached higher levels. Then, it arises that *Q. robur* extract probably induced a faster and higher production of ROS than did *P. australis* extract. A probable reason for this observation could be because *Q. robur* has high phenolic contents (Salminen et. al., 2004) whose redox properties allow them to act as reducing agents, hydrogen donors and singlet oxygen quenchers (Rive-Evans et. al., 1995). POD and GPx catalyses the breakdown of H_2O_2 produced during stress (Reviewed by Arora et. al., 2002). It is apparent that there was a progressively high production of ROS from the onset of stress due to exposure to the leaf extracts and this was accompanied by a concomitant increase in the activity of the antioxidant enzymes. As the enzymes expedite their removal, ROS levels begin to decrease over time (after 48 h) until they are eventually maintained at ambient levels. Previous control levels, both in terms of photosynthetic efficiency and antioxidative enzymes, were restored regarding *P. australis* extracts. With respect to *C. demersum* exposed *to Q. robur* extracts, it turns out that the effects on the enzymes and photosynthesis did not change between 48 and 168 h (apart from mGST), neither did they return to control levels. This suggests on the one hand that more time than covered in this study was probably required for complete recovery from the effects of *Q. robur* extracts. On the other hand, an alternative stable-state, different from the previous one, may have been established (Tausz et al., 2004) since no change in photosynthesis or enzyme activity occurred between 48 and 168 h. In this probable alternative steady-state, *C. demersum* may utilize alternative protection mechanisms such as photorespiration, light dissipation (Tausz et. al., 2004) and energetic flexibility (Dobrota, 2006) to endure the potentially toxic effects of *Q. robur* extracts. However, photorespiration as an alternative survival strategy may be counterproductive since it is also a means by which H_2O_2 is produced (Dat et. al., 2000).

Whereas very high levels of ROS can lead to phytotoxicity, relatively low levels can be used for acclimatory signalling. Previous studies as reviewed in Dat et. al. (2000) seems to suggest that exposure of plants to mild oxidative stress confers some measure of resilience to further stress. This review corroborates the correlation between antioxidant potential and stress tolerance, indicating that acclimation to elevated ROS is associated with enhanced activity of antioxidant enzymes. In the present study, it is likely that within 24 h an increase in ROS levels may have evoked an initial increase in the antioxidant enzymes thereby conferring 'immunity' to subsequent ROS build-up. In spite of the damaging effects often associated with ROS, they play a significant role as metabolic signals to initiate defense mechanisms that enhance tolerance to potentially toxic effects (Noctor, 2006). As a result, when plants are challenged with high ROS levels, the enzymatic machinery is activated and this causes an eventual reduction in ROS levels (Blokhina et al., 2003). In the present study, increase in the activity of antioxidative defense enzymes (GST, POD, GR, GPx) reached maximum after 24 h of exposure to both leaf extracts, except cGST (4-HNE substrate) where highest activity was attained at 12 h exposure to *P. australis* extract. Subsequently, enzyme activities declined rather sharply within 48 h of exposure to an environmentally relevant DOC concentration suggesting an onset of acclimation to stress imposed by the leaf extracts. In a related study, Klenell et al. (2005) demonstrated a link between acclimation of the *Arabidopsis* mutant (Chaos) to photo-oxidative stress with lower ascorbate levels, less induction of ascorbate peroxidases and lower H_2O_2 production. At the same time, the regulation of the CAO gene (coding for the chloroplast recognition particle cpSRP43) was found to be inhibited during acclimation to high light and chilling temperatures. The changes in activity and the activation state of the defense enzymes in the present study are consistent with short-term (24 h) responses to other humic or humic-like substances observed in previous studies (Pflugmacher et. al., 1999, 2003). Aguilera et. al. (2002) also found similar acclimational responses of ROS-scavenging enzymes in marine macroalgae triggered after 84 h exposure to ultraviolet radiation. Vranova et al. (2002) observed an increased expression of GPx gene in ROS-acclimated tobacco leaves and suggested that this protein plays a role in the enhancement of tolerance to oxidative stress. It appears that the photosynthetic machinery and the oxidative defense system in *C. demersum* are more resilient to the effects caused by *P. australis* extract than by *Q. robur* extract. It is probable that the delayed recovery of photosynthesis from the effect of *Q. robur* extract was, in addition to its chemical properties (Salminen et. al., 2004), a compensation for the corresponding delay in the onset of establishing its effects. Similar to photosynthesis, the effect of *Q. robur* extracts on enzymatic response, apart from GR, probably required more time for complete recovery than the period covered in this study. Chloroplasts, the site of photosynthesis, are particularly at risk during stress because of the relatively higher production of ROS in this organelle during normal photosynthesis (Alscher, 2002). Superoxide ($O_2^{\cdot-}$) radicals produced during photosynthesis are dismutated by superoxide dismutase (SOD) enzymes to H_2O_2. The increase in activity of the enzymes (POD and GPx) involved in the removal of H_2O_2 observed in this study, contributes to maintenance of low ROS levels in the chloroplast. Consequently, it was shown that as the activity of the ROS-scavenging enzymes decreased, which is an indication of decreasing ROS levels, photosynthetic recovery was initiated. Furthermore, both the antioxidative system and photosynthetic oxygen production in *C. demersum* were in parallel resilient to *P. australis* extract during 1 week of exposure. *C. demersum* was only partially resilient to *Q. robur* extract within the period investigated. It thus seems evident that an increase in activity of antioxidative enzymes is, at least in part, a means whereby *C. demersum* can acclimate to oxidative stress imposed by leaf extracts.

In conclusion, presented data demonstrate the ability of *Ceratophyllum demersum* to acclimate and adjust to stress originating from exposure to *P. australis* extract. The antioxidative system in *C. demersum* responded relatively sooner in suppressing stress due to *Q. robur* leaf extracts. The stress posed on the photosynthetic system of *C. demersum* was completely overcomed within one week of exposure to *P. australis* extract, but persisted throughout the experiment in the case of *Q. robur* extract. Similar to photosynthesis, the enzyme system reached maximum within 24 h followed by a complete recovery for *C. demersum* in *P. australis* extract and partial recovery for *C. demersum* in *Q. robur* extracts within the period investigated. These findings thus underscore the need to carefully set (long enough) pre-acclimation period prior to toxicity tests in order to cover the entire period required for complete recovery so as to exclude confounding effects. Overall, this study indicates that the ecological success or resilience of macrophytes (e.g. *C. demersum*) in aquatic systems is, at least in part, supported by an enhanced ROS-scavenging system, allowing fast acclimation, depending on the stress factor considered, to changes in environmental conditions.

4.1.2 Molecular and antioxidant response experiments

4.1.2.1 Experiment 3: Regulation of glutathione reductase gene expression and glutathione redox dynamics in *C. demersum* by leaf litter-derived DOC

Given the important role of GR in maintaining glutathione redox balance in plant cells, coupled with the fact that the gene encoding this enzyme has not been sequenced from relevant aquatic species, it was of interest to isolate the sequence and evaluate the expression of the gene encoding this enzyme. The modulation of GR expression and glutathione redox dynamics in *C. demersum* due to leaf extracts is discussed in the context of providing protection against oxidative stress.

We showed that during the stress conditions of increased DOC concentrations, H_2O_2 levels in *C. demersum* increased. At the same time, glutathione use increased, as evidenced by the lower GSH/GSSG ratio. These are similar findings to other stress conditions such as Cu toxicity (Rijstenbil et al., 1998) and light intensity (Irihimovitch and Shapira 2000). The increased H_2O_2 production in the present study is likely a consequence of the aromaticity of the DOC pool which probably reflects the presence of quinones, tannins and phenolic compounds in crude leaf extracts (Salminen et al, 2004). The role of quinone moities in driving DOC mediated redox reactions (Fimmen et al, 2007) can be largely attributed to the formation of semiquinone radicals that donate electrons to molecular oxygen, forming superoxide anions (Weir et al, 2004). These anions can then be further converted by superoxide dismutase (SOD) enzymes to form H_2O_2 (Blokhina et al, 2003). The levelling off of H_2O_2 at the highest tested DOC level in both extracts may be due to an overwhelming of the SOD enzyme activity such that higher stress level does not lead to further conversion of superoxide radicals to form H_2O_2.

The lower GSH/GSSG ratio manifested by *C. demersum* in the presence of leaf extracts is likely due to the increased levels of mGST, cGST and GPx under stress conditions (Kamara and Pflugmacher, 2007a). In addition, the antioxidant role of GSH is based on its free-radical scavenging capacity, often resulting in its conversion to GSSG. Consistent with our previous report (Kamara and Pflugmacher, 2007a), oak extracts imposed more stress on *C. demersum* than reed extracts as evidenced by mostly higher H_2O_2 levels and lower GSH/GSSG ratios. Thus *C. demersum* was able to acclimate faster (Kamara and Pflugmacher, 2007b) to reed-induced stress as indicated by increased GSH/GSSG ratio following reduced GSH utilization after 24 and 48 h exposure. This may be explained, at least in part, by the higher aromaticity and HS measured in oak compared with reed extracts, since the effect of plant leachates in terms of uptake and oxidative stress induction depends on the relative proportion and size of DOC fractions and aromaticity (Wang et al., 1999, Nardi et al, 2002). In fact, *Q. robur* (oak) leaves are known to consist of high amounts of tannins, including the rare dimeric ellagitannin, cocciferon D2, and phenolic compounds (Salminen et al, 2004). Oxidation of these compounds can thus generate higher amounts of superoxide anion radicals and ultimately H_2O_2 via conversion by SOD.

An overall increase in total glutathione (GSH plus GSSG) content observed in concomitance with increased H_2O_2 level indicates a leading role of glutathione in the adaptive response of *C. demersum* to leaf extract-induced stress. Earlier studies have also demonstrated an increase in glutathione concentrations in relation to H_2O_2 accumulation (Vanacker et al, 2000) and sometimes accompanied by a modulation of the genes encoding the enzymes of the GSH biosynthetic pathway (Xiang and Oliver, 1998).

A modulation of GR expression due to DOC from oak leaf extract was manifested in the present study. The higher proportion of humic-like substances and aromaticity in oak extracts than reed extracts suggest humic-like substances might trigger the up-regulation of GR, which ultimately enhances the re-conversion of GSSG to GSH required for various ROS-scavenging and detoxification mechanisms during oxidative stress (Romero-Puertas et al., 2006). Induced GR expression in response to oak leaf extract-induced stress was generally short-lived and did not persist beyond 24 h, suggesting fast acclimation (Kamara and Pflugmacher, 2007b) at the molecular level. The reason for the observed reed extract-induced GR repression after 4 h is unclear but could be due to post-translational changes during the early steps of GR protein synthesis.

The leaf extract-induced stress, together with augmented GR activity as previously reported (Kamara and Pflugmacher, 2007a), in the absence of parallel enhancement of GR expression (except at 10 mg L^{-1} *Q. robur* extract), indicates that post-translational activation of this enzyme is likely. Chemical (e.g phosphorylation) and structural (e.g disulphide bridge formation) changes are typical post-translational modifications that could lead to alteration of GR protein activity (Montoya-García et al, 2002; Bae and Sicher, 2004). These results are in agreement with those reported by Lara-Nunez et. al. (2006) who also found no correlation between enzyme activity and mRNA expression of GR, among other antioxidant enzymes, under allelochemical stress conditions.

Overall, it appears that GR in *C. demersum* is an early response gene, depending on the stress factor considered. The response was manifested within 24 h at the latest. This might explain why, among other reasons (Kamara and Pflugmacher, 2007b), *C. demersum* is intrinsically tolerant to oxidative stress posed by DOC from dead and decaying plant litter and thus able to survive in its natural habitat. That these responses occurred at an environmentally relevant DOC concentration (10 mg L^{-1}) suggests that GR could serve as an important biomarker gene for non-synthetic natural compounds in aquatic environments. In general, these findings suggest that leaf litter-derived DOC plays an important role as chemical signal at the molecular and cellular levels in shallow freshwaters, as well as demonstrating a clear connectivity between terrestrial and aquatic ecosystems.

4.2 Discussion: Experiments with *Lemna minor*

4.2.1 Biochemical responses of *L. minor* to leaf litter-derived DOC

4.2.1.1 Experiment 4: Short-term (24 h) biochemical dose-responses

In these experiments, we investigated the likelihood of cell damage by lipid peroxidation and the protective capacity of GSH and antioxidative enzymes. As mentioned earlier, aqueous leaf litter extracts are known to consist of tannins and quinones (Salminen et al., 2004) whose derivatives (semiquinone radicals) can lead to the formation of superoxide anions by donating electrons to oxygen (Weir et al., 2004). The increased H_2O_2 content exhibited by *L. minor* in the presence of aqueous leaf extracts is likely a consequence of the observed increased level of SOD under stress conditions, converting superoxide anion radicals to H_2O_2. However, H_2O_2 also is a toxic reactive oxygen species and its accumulation could lead to oxidative stress. In response to increased H_2O_2 levels, POD, CAT and GPx enzymes, all of which catalyze the breakdown of H_2O_2 to harmless water molecules, were concurrently activated. Increase in antioxidant enzyme activities in response to elevated H_2O_2 contents due to other stress factors have been shown in other studies using *L. gibba* (Babu et al., 2005)

The biotransformation system involving both cGST and mGST were modulated by DOC from the two leaf extracts irrespective of the substrate used for the enzymatic assay. The increased activity of cGST assayed with 4-HNE as a substrate suggests active conjugation of LPO products and protection from further oxidative damage (Yang et al., 2003). GSH is a common non-enzymatic antioxidant involved in direct quenching of ROS and as a cofactor in enzymatic defense. The elevation of GPx as well as m- and cGST meant an increased utilization of GSH leading to the formation of the so-called glutathione-conjugates. This increased use of GSH led to a depletion of the GSH-pool and a disruption of the glutathione redox balance as evidenced by a decrease in total glutathione and GSH/GSSG ratio, respectively, at high DOC levels. The enhanced activity of GR at high DOC levels indicates a leading role of GR in the maintenance and resuscitation of the glutathione redox balance by reducing GSSG to GSH.

Increase in LPO in *L. minor* due to elevated ROS formation in the presence of other stress factors have been previously reported (Razinger et al., 2007; Tkalec et al., 2007). In the present study, exposure of *L. minor* to both leaf extracts led to increase in H_2O_2 formation.

Due to the elevated level of H_2O_2 and probably other ROS, cellular damage in the form of LPO could be expected. On the other hand, concomitant activation of antioxidant and biotransformation enzymes resulted in protection against oxidative damage as shown by similar MDA levels in exposed and control plants. However, it appears that the antioxidative defense capacity in *L. minor* was overwhelmed at the highest reed DOC (100 mg L^{-1}) level tested, leading to significant ($P < 0.05$) LPO.

In summary, we have demonstrated that leaf litter-derived DOC potently augments the generation of H_2O_2 with a consequent induction of antioxidative defense enzymes, accompanied by reduction of GSH content in *L. minor*. As a consequence, LPO was averted, with the exception of highest reed-derived DOC. Cellular damage noted in *P. australis*-exposed *Lemna* was within the range of DOC (1 – 100 mg L^{-1}) concentration often measured in oligotrophic freshwater ecosystems (Wetzel, 2001), suggesting that *Lemna* may be susceptible to high *P. australis*-derived DOC especially in the littoral zones of shallow lakes, often dominated by reeds.

4.2.1.2 Experiment 5: Acclimatory (time) responses of *L. minor* to the effects of leaf litter-derived DOC

Phenotypic plasticity manifested by *L. minor* in terms of biotransformation and antioxidative enzyme responses in the two leaf extracts were variable. In *P. australis* extract, the enzymatic responses were mostly late with the exception of GR where induction occurred after 1 h of exposure. GPx and cGST (CDNB) were not induced until after 24 h of exposure to *P. australis* extract. In the case of *Q. robur* extract, inductive responses were relatively sooner, ranging from 4 h (GR, cGST-CDNB) to 12 h (GR, cGST-4HNE). The start of induction of the various antioxidative enzymes is an indication of the onset of oxidative stress probably as a result of increased ROS production due to leaf extracts exposure. If so, it then arises that *Q. robur* leaf extract likely induced an earlier production of ROS in *L. minor* than *P. australis* leaf extract. Therefore, decrease in enzyme activity that will ultimately restore ambient enzyme levels during prolonged exposure to leaf extracts would imply an adaptive or acclimatory response (Kamara and Pflugmacher, 2007b). In *L. minor*, all enzymes (except GR in *Q. robur* extract) of the oxidative defense system, including the biotransformation enzyme GST, did not return to control levels in both extracts. This suggests that *L. minor* was, within the investigated period, unable to acclimate to the effects of the leaf extracts and thus keeping the defense enzymes constantly activated in order to prevent or minimize oxidative damage resulting from continued high ROS generation. Our results further showed that most of the enzymes investigated, notably CAT, POD, cGST and GR, displayed biphasic responses depending on the stress factor (*Q. robur* or *P. australis* extract) and enzyme substrate (CDNB or 4HNE). Biphasic curves are characteristic of stress responses (Teisseire and Guy, 2000; Lichtenthaler, 1996). During one week of exposure to leaf extracts, maxima of enzyme induction were attained between 12 and 48 h and then activities decreased slightly compared with preceding enzyme levels, but increased again afterwards. These points of inflexion suggest an unsuccessful attempt to acclimate to the stress conditions. Prolonged exposure often leads to hardening of the plant with the result of a re-establishment of previous ambient enzyme levels or a new alternative stable state (Tausz et al., 2004). This was however not the case for *L. minor* after 7 days of exposure to both leaf extracts, suggesting that this macrophyte is less resilient and probably requires more time than covered in this study to acclimate to the leaf extracts-induced stress effects. Therefore, longer exposure periods are necessary to be able to make a strong statement about the physiological acclimatory potential or susceptibility of *L. minor* to the leaf extracts tested.

4.2.2 *Lemna minor* Bioassay Experiments

4.2.2.1 Experiment 6: Photosynthesis and growth responses of *L. minor* L.

L. minor as a sensitive surface-floating aquatic macrophyte species, with a standardized testing for toxicity of chemical substances (OECD 2006, draft 221), was used to test the potential toxic effects of leaf extracts on growth and photosynthesis. Chlorophyll pigment contents were also determined to evaluate if photosynthetic efficiency was influenced by pigments concentration in a way that is indicative of the effect of leaf extracts.

Changes in chlorophyll pigments have been frequently used as a sensitive physiological endpoint for the assessment of biological effects of substances coming in contact with plants (Turgut and Fomin, 2002; Viet and Moser, 2003; Brain et al., 2004; Cedergreen et al., 2007). Moreover, pigment contents are among the endpoints recommended in the standard OECD guidelines for *Lemna* toxicity bioassays (OECD 2006). Our results indicate that both *Q. robur* and *P. australis* extracts influenced chlorophyll pigments, with *Lemna* showing higher sensitivity to *P. australis* than to *Q. robur* extract. Both extracts resulted in total chlorophyll pigment reduction at high DOC concentration (100 mg L^{-1}) within the first 3 days of exposure. At the end of the experiment (day 7), *P. australis*-induced chlorophyll reduction occurred at low DOC concentration (1 mg L^{-1}) as well, perhaps as a consequence of inhibited pigment biosynthesis. These results are consistent with findings from Everall and Lees (1996) who observed significant decline in cyanobacterial chlorophyll *a* levels in a reservoir treated with 50 g m^{-3} barley straw. Chlorophyll *a/b* ratio significantly declined at higher DOC levels in both extracts, indicating that chlorophyll *a* was more affected than chlorophyll *b* and suggest a stress condition in *L. minor*. These are similar findings to other stress conditions such as Cd toxicity (Li et al., 2008) and organic matter (Pflugmacher et al., 1999).

Photosynthetic rates are closely related to the content of light harvesting pigments. In both extracts, reduction in pigment contents at higher DOC concentrations coincided with a decrease in photosynthetic oxygen release, showing that photosynthesis inhibition was also a consequence of chlorophyll reduction. In *Q. robur* extract, however, although chlorophyll contents were not affected at lower DOC levels (1 and 0.1 mg L^{-1}) after 7 days, photosynthesis was significantly inhibited, suggesting damage to other parts of the photosynthetic system (e.g. electron transport chain) (Weir et al., 2004; Pflugmacher et al., 2006).

Clear detrimental effects were observed on all measured growth endpoints with varying sensitivities. Whereas frond number-specific growth rate was not significantly inhibited at 0.1 mg L^{-1} DOC in *P. australis* extract, a significantly lower number of fronds compared with control value was recorded. This suggests that growth rate is a more conservative parameter than changes in absolute frond numbers.

A low EC_{50} is often considered as indicative of a process being affected more strongly. Therefore, differences in EC_{50} between endpoints reflect the relative sensitivity of *L. minor* to the two extracts. In both extracts, lower EC_{50}s were calculated for dry weight (DW) than for frond number (FN). This is in agreement with report by Cleuvers and Ratte (2002), showing that dry weight was a more sensitive endpoint in detecting growth inhibition. In general, the two extracts seem to have different mode of action on the growth of *L. minor*, and this appears to be concentration dependent. At lower levels of DOC exposure, *L. minor* was more sensitive to *Q. robur* than to *P. australis* extract as indicated by lower DW-specific LOEC and NOEC values. Considering FN-specific growth rates, identical LOEC and NOEC values suggest that the multiplication rate of *L. minor* was affected in a similar way by the two extracts at low DOC levels. Similarly, earlier studies have demonstrated different growth responses manifested by algae (Matsunaga et al., 1999; Ridge et al., 1999) and fungi (Bärlocher, 1992;

Gryndler et al., 2005) to different DOC sources. In the present study, it was evident that growth inhibition was a function of DOC concentration. In the two extracts, DOC concentrations that caused DW-specific growth reduction coincided with photosynthetic inhibition after 7 days of exposure, demonstrating that photosynthetic inhibition is one mechanism by which leaf extracts influence growth rates of *L. minor*. In the present study, pigments were, in most cases, less sensitive to the leaf extracts than growth endpoints. Nevertheless, they served as complement to the growth assay because they demonstrate damage to processes other than cell division (Greenberg et al., 1992). In conclusion, data presented shows that physiological (pigments and photosynthesis) responses largely correspond to growth effects in *L. minor* exposed to *Q. robur* and *P. australis* leaf litter extracts. Modulation of *Lemna* growth as indicated by reduction in frond numbers and biomass production clearly demonstrate the potential of *Q. robur* and *P. australis* extracts to influence aquatic macrophyte dynamics and hence to structure aquatic ecosystems.

4.3 Discussion: Overview of the physiological responses of *C. demersum* and *L. minor* to leaf litter-derived DOC

Oxidative stress responses were evident in both macrophytes (*C. demersum* and *L. minor*) after short term (24 h) exposure to oak and reed leaf litter extracts. The antioxidative enzyme responses were proportionately more pronounced in *C. demersum* than in *L. minor* when compared with their respective controls. This suggests that *C. demersum* was more stressed *than L. minor*, probably due to its submersed nature, allowing more surface area contact with the leaf extracts. The two macrophytes exhibited similar non-enzymatic glutathione-mediated antioxidative mechanisms in response to DOC stress. In both macrophytes, total glutathione and GSH/GSSG ratio declined at high DOC levels of exposure, accompanied by a concomitant increase in GR and other antioxidative and biotransformation enzyme activity. In both macrophytes, the high response of the antioxidative system was sufficient to avert cellular damage (lipid peroxidation), except at the highest tested DOC concentration of reed extracts where *L. minor* became overwhelmed and lipid peroxidation became evident.

In terms of prolonged (168 h) sub-acute exposure of the two macrophytes to the leaf litter extracts, it turns out that *C. demersum* was able to acclimate to reed extracts within the period investigated. Although no complete acclimation was observed in *C. demersum* exposed to oak extracts, a tendency of decreasing enzyme levels was already initiated, suggesting the possibility of recovery through time or an establishment of a new alternative stable state

(Tausz et al., 2004). In the case of *L. minor*, biochemical acclimation was not detectable within the period investigated. It thus seems that *C. demersum*, by virtue of its vulnerability due to greater surface area contact with the leaf extracts, has evolved an ability to acclimate faster to DOC-derived stress, depending on the source of DOC. It can be concluded that the effects of leaf extracts on the biochemical pathways and the accompanying antioxidative enzymatic adjustment of the macrophytes is species specific and depends also on the stress factor considered.

Photosynthesis was inhibited in both macrophytes, probably with differing modes of action. In *C. demersum*, photosynthetic inhibition was due mainly to the hypothesized electron transport disruption system because no effects were observed on the chlorophyll pigments. In *L. minor*, photosynthetic inhibition was, at least in part, due to chlorophyll pigment reduction.

5. Conclusions and perspectives

It was clear that leaf litter-derived DOC affects the macrophytes by inhibiting the photosynthetic process. The mechanism of action is not clear. In *C. demersum*, chlorophyll pigments were not affected, yet significant photosynthetic inhibition occurred, suggesting that the mechanism of action was mainly that proposed by Pflugmacher et al. (2006) involving the interruption of the electron transport chain. In the case of *L. minor*, both pigment reduction and photosystem II inhibition may have been involved since significant reduction in chlorophyll pigments was also observed.

Within 24 h, environmentally relevant levels of DOC from oak and reed leaf litter caused significant stimulation of biotransformation and antioxidative enzymes. When *L. minor* was exposed to the same DOC concentrations for 7 days, chlorophyll pigments decreased, photosynthesis and growth were inhibited. Thus the onset of antioxidative enzyme responses generally occurred before the beginning of visible growth inhibitory effects in *L. minor*. This suggests that modulation of biotransformation and antioxidative enzyme activities could constitute an early indication of DOC 'toxicity', thereby supporting the hypothesis that antioxidative enzymes principally play a signalling role and hence may be useful as biomarkers for the early detection of ecological disturbances.

At the molecular level, GR gene was overexpressed at a DOC concentration of 10 mg L^{-1} in oak but not in reed extracts. This was in contrast to GR activity which was significantly induced in both leaf extracts at DOC levels ranging from 0.1 to 100 mg L^{-1}. This implies that increases in the activity of GR in response to external stressors are not necessarily reflected in changes of the steady-state levels of GR protein or mRNA.

In general, results obtained for the different biochemical parameters showed mostly a good correlation and indicating a good functioning of the enzymatic and non-enzymatic (GSH) defense mechanism in both macrophytes. The activation of the enzymatic and non-enzymatic antioxidant system was sufficient to prevent lipid peroxidation except at the highest (100 mg L^{-1}) reed DOC level tested where *L. minor* became susceptible. It was thus clear that the antioxidative system served as a signal response to oxidative stress, acting in a concerted manner to avert cellular damage.

The adaptive or acclimatory responses of the two macrophytes to the leaf extracts was species specific and depends on the source of DOC. Whereas *C. demersum* acclimated within one week to reed extracts, *L. minor* remained stressed and in some cases manifested biphasic enzyme responses due to both extracts throughout the period investigated.

Taken together, the present study shows that aquatic and terrestrially-derived leaf-litter DOC has the potential to impact aquatic macrophytes by imposing oxidative stress, slowing down photosynthesis and reducing growth rate. Therefore, we suggest that both allochthonous (oak) and autochthonous (reed) leaf litter has the potential to structure aquatic ecosystems. We use the term 'potential' because an ecological field situation constitutes a more complex matrix and are largely different from a DOC-treated laboratory experiment. For instance, we have to take into account the remote possibility that other substances (e.g. heavy metals, Laing et al., 2006), if taken up and accumulated, may be present in crude aqueous leaf litter extracts. Consequently, possible interactions with synthetic pollutants, modification of bioavailability, microbial- and photo-modification of leaf extracts, etc. could modify the observed effects if tested in the field. Hence, results obtained in the present study should be interpreted with caution in the context of macrophyte dynamics in natural ecosystems. Nevertheless, the relatively strong effects from the two leaf litter extracts suggest that these phenomena could occur in *P. australis*-dominated shallow lakes or aquatic systems with a *Q. robur*-dominated catchment area.

Although this study revealed important macrophyte responses at the molecular, cellular and organismal levels, it remains to be shown that some specific compounds in the crude leaf extracts provoked the observed effects. For this, further studies would be required that will combine a detailed characterisation of the leaf extracts and pairs it with various phenotypic responses in order to have a better understanding of the mechanism of action. Bioassay-guided fractionation of the leaf extracts could ultimately lead to the establishment of a cause-effect relationship between compounds in the leaf extracts and the macrophyte responses.

6. References

Aguilera, J., Dummermuth, A., Karsten, U., Schriek, R. and Wiencke, C. 2002. Enzymatic defences against photooxidative stress induced by ultraviolet radiation in Arctic marine macroalgae. Polar biology 25, 432-441.

Alin, P., Danielson, U.H. and Mannervik, B. 1984. 4-hadroxyalk-2-enals are substrates for glutathione transferase. FEBS letters 179, 267-270.

Alscher, G.R., Erturk, N. and Heath, S.H. 2002. Role of superoxide dismutase (SODs) in controlling oxidative stress in plants. Journal of Experimental Botany 53, 1331-1341.

Anderson, M.E. 1985. Determination of glutathione and glutathione disulphide in biological samples. Methods in Enzymology 113, 548-555.

Anderson, M.D., Prasad T.K. and Stewart, C.R. 1995. Changes in isozyme profiles of catalase, peroxidase and glutathione reductase during acclimation to chilling in mesocotyls of maize seedlings. Plant Physiology 109, 1247-1257.

Arora, A., Sairam, R.K. and Srivastava, G.C. 2002. Oxidative stress and antioxidative system in plants. A review: Current Science 82, 1227-1238.

Babu, T.S., Tripuranthakam, S. and Greenberg, B.M. 2005. Biochemical responses of the aquatic higher plant Lemna gibba to a mixture of copper and 1,2.dihydroxyanthraquinone: Synergistic toxicity via reactive oxygen species. Environmental Toxicology and Chemistry 24, 3030-3036.

Bae, H. and Sicher, R. 2004. Changes of soluble protein expression and leaf metabolite levels in Arabidopsis thaliana grown in elevated atmospheric carbon dioxide. Field Crops Research 90, 61-73.

Baier, M and Dietz, K.J. 2005. Chloroplasts as source and target of cellular redox regulation: a discussion on chloroplast chloroplast redox signals in the context of plant physiology. Journal of Experimental Botany 56, 1449-1462.

Bärlocher, F. 1992. Effects of drying and freezing autumn leaves on leaching and colonization by aquatic hyphomycetes. Freshwater Biology 28, 1-7.

Baudhuin, P.H., Beaufay, Y., Rahman-Li, Y., Sellinger, O.H., Watliaux, R., Jacques, P. and Duve C. 1964. Tissue fractionation studies XVII. Intracellular distribution of monoamine oxidase, aspartate amino-transferase, d-amino acid oxidase and catalase in rat liver tissue. Biochemical Journal 92, 179-187.

Bechtold, U., Karpinski, S. and Mullineaux, P.M. 2005. The influence of the light environment and photosynthesis on oxidative signaling responses in plant-biotrophic pathogen interactions. Plant Cell and Environment 28, 1046-1055.

Bergmeyer, H.U. (ed.) (1983-86) Methods of enzymatic analysis. Vol. I-XII, 648-653. VCH, Weinhein, Deutschland.

Blodau, C., Basiliko, N. and Moore, T.R. 2004. Carbon turnover in peatland mesocosms exposed to different water table levels. Biogeochemistry 67, 331-351.

Blokhina,O., Virolainen, E. and Fagerstedt, K.V. 2003. Antioxidants, oxidative damage and oxygen deprivation stress: a review. Annals of Botany 91, 179-194.

Botsglou, N.A., Fletouris, D.J., Papageorgiou, G.E., Vassilopoulos, V.N., Mantis, A.J. and Trakatellis, A.G. 1994. Rapid, sensitive, and specific thiobarbituric acid method for measuring lipid peroxidation in animal tissue, food, and feedstuff samples. Journal of Agriculture and Food Chemistry 42, 1931-1937

Bradford, M.M. 1976. A rapid and sensitive method for the quantification of microgram quantities of protein utilizing the principle of protein dye-binding. Analytical Biochemistry 72, 248–254

Brain, R.A., Johnson, D.J., Richards, S.M., Hanson, M.L., Sanderson, H., Lam, M.W., Young, C., Mabury, S.A., P.K. Sibley, P.K. and Solomon, K.R. 2004. Microcosm evaluation of the effects of an eight pharmaceutical mixture to the aquatic macrophytes *Lemna gibba* and *Myriophyllum sibricum*. Aquatic Toxicology 70, 23-40.

Brooks, P.D., McKnight D.M., Bencala, K.E., 1999. The relationship between soil heterotrophic activity, soil dissolved organic carbon (DOC) leachate, and catchment-scale DOC export in headwater catchments. Water Resources Research 35, 1895-1902.

Byl, T.D., Sutton, H.D.and Klaine, S.J. 1994. Evaluation of peroxidase as a biochemical indicator of toxic chemical exposure in the aquatic plant Hydrilla verticillata. Royle. Environmental Toxicology and Chemistry 13, 509-515.

Carlberg, I. and Mannervik, B. 1985. Glutathione reductase. Methods in Enzymology 113, 484-490.

Castillo, F.J., Miller, P.R. and Greppin, H. 1987. Extracellular biochemical markers of photochemical oxidant air pollution damage to Norway spruce. Experimentia 43: 111-115.

Cedergreen, N., Abbaspoor, M., Sorensen, H. and Streibig, J.C. 2007. Is mixture toxicity measured on a biomarker indicative of what happens on a population level? A study with *Lemna minor*. Ecotoxicology and Environmental Safety 67, 323-332.

Cleuvers, M. and Ratte, H.T. 2002. Phytotoxicity of coloured substances: is *Lemna* Duckweed an alternative to the algal growth inhibition test? Chemosphere 49, 9-15.

Cole, J.J., Pace, M.L., Carpenter, S.R. and Kitchell, J.F., 2000. Persistence of net heterotrophy in lakes during nutrient addition and food web manipulations. Limnology and Oceanography 45, 1718-1730.

Dat, J., Vandenabeele, E., Vranova, M., Van Montagu, M., Inze, D. and Van Breusegem, F. 2000. Dual action of the active oxygen species during plant stress responses. Cellular and Molecular Life Sciences 57, 779-795.

DIN EN 1484, 1998. Anleitung zur Bestimmung des gesamten organischen Kohlenstoffs (TOC) und des gelösten organischen Kohlenstoffs, DEV, H3, 40. Lieferung.

Dobrota, C. 2006. Energy dependent plant stress acclimation. Review of Environmental Biotechnology 5, 243-251.

Dröge, W. 2002. Free radicals in the physiological control of cell function. Physiological Reviews 82, 47-95.

Drotar, A., Phelps P. and Fall R. 1985. Evidence for glutathione peroxidase activities in cultured plant cells. Plant Science 42, 35-40.

Evans C.D., Monteith D.T., Cooper DM (2005) Long-term increases in surface water dissolved organic carbon: Observations, possible causes and environmental impacts. Environmental Pollution 137, 55-71.

Everall, N.C. and Lees, D.R. 1996. The use of barley-straw to control general and blue-green algal growth in a Derbyshire reservoir. Water Research 30, 269-276.

Fimmen, R.L., Cory, R.M, Chin, Y.-P, Trouts, T.D. and McKnight, D.M. 2007. Probing the oxidation-reduction properties of terrestrially and microbially derived dissolved organic matter. Geochimica et Cosmochimica Acta 71, 3003-3015.

Foyer, C.H., Lescure, J.C., Lefebvre, C., Morot-Guadry, J.F., Vincent, M. and Vaucheret, H. 1994. Adaptations of photosynthetic electron transport, carbon assimilation and carbon partitioning in transgenic *Nicotiana plumbaginifolia* plants to changes in nitrate reductase activity. Plant Physiology 104, 171-178.

Foyer, C.H. and Noctor, G. 2005. Redox homeostasis and antioxidant signaling: a metabolic interface between stress perception and physiological responses. Plant cell 17, 1866-1875.

Fridovich, I. 1998.Oxygen toxicity: a radical explanation. Journal of Experimental Biology 201, 1203-1209.

Gessner, M.O. 2000. Breakdown and nutrient dynamics of submerged *Phragmites* shoots in the littoral zone of a hardwater lake. Aquatic Botany 66, 9-20.

Greenberg, B.M., Huang, X.D., Dixon, D.G., 1992. Applications of the aquatic higher plant *Lemna gibba* for ecotoxicological risk assessment. J. Aquat. Ecosystem Health 1, 147-155.

Grimm, N.B., Gergel, S.E., McDowell, W.H., Boyer, E.W., Dent, C.L., Groffman, P., Hart, S.C., Harvey, J., Johnston, C., Mayorga, E., McClain, M.E. and Pinay, G. 2003. Merging aquatic and terrestrial perspectives of nutrient biogeochemistry. Oecologia 137, 485–501.

Grundhöfer, P., Niemetz, R., Schilling, G., Gross, G.G., 2001. Biosynthesis and subcellular distribution of hydrolyzable tannins. Phytochemistry 57, 915-927.

Gryndler, M., Hršelová, H., Sudová, R., Gryndlerová, H., Řezáčová, V. and Merhautová, V. 2005. Hyphal growth and mycorrhiza formation by the arbuscular mycorrhizal fungus *Glomus claroideum* BEG 23 is stimulated by humic substances. Mycorrhiza 15, 483-488.

Habig, W., Pabst, M.J. and Jakoby, W.B. 1974. Glutathione S-transferase: the first step in mercapturic acid formation. Journal of Biological Chemistry 249, 7130-7139.

Hausladen, A. and Alscher, R.G. 1993. Glutathione. In: Alscher, R.G. and Hess, J.L., (eds.). Antioxidants in higher plants CRC press, Inc., pp. 1-30.

Herschbach, C. and Rennenberg, H. 1994. Infuence of glutathione (GSH) on net uptake of sulphate and sulphate transport in tobacco plants. Journal of Experimental Botany 45, 1068-1076.

Hieber, M. and Gessner, M.O. 2002. Contribution of stream Detrivores, Fungi and Bacteria to leaf breakdown based on biomass estimates. Ecology 83, 1023-1038.

Hollis, G.E., Jones, T.A., 1991. Europe and the Mediterranean Basin. In: Finlayson, M. and Moser, M. (Eds.). Wetlands, pp. 27-56.

Huber, S.A. and Frimmel, F.H. 1996. Liquid chromatography and organic carbon detection (LC-OCD): A fast and reliable method for the characterization of hydrophilic organic matter in natural waters. *Vom Wasser* 86, 277-290.

Imlay, J.A. and Linn, S. 1986. DNA damage and oxygen radical toxicity. Science 240, 1302-1309.

Inskeep, W.P. and Bloom, P.R. 1985. Extinction coefficients of chlorophyll a and b in N,N-dimethylformamide and 80% of acetone. Plant Physiology 77, 483-485.

Irihimovitch, V. and Shapira, M. 2000. Glutathione redox potential modulated by reactive oxygen species regulates translation of Rubisco large subunit in the chloroplast. Journal of Biological Chemistry 275, 16289-16295

Jones, R.I. 2005. Limnology of humic waters: special theme or universal framework? Verhandlungen der Internationalen Vereinigung für theoretische und angewandte Limnologie 29, 51-60.

Kamara, S. and Pflugmacher, S., 2007a. *Phragmites australis* and *Quercus robur* leaf extracts affect antioxidative system and photosynthesis of *Ceratophyllum demersum*. Ecotoxicology and Environmental Safety 67, 240-246.

Kamara, S. and Pflugmacher, S. 2007b. Acclimation of *Ceratophyllum demersum* to oxidative stress imposed by *Phragmites australis* and *Quercus robur* leaf extracts. Ecotoxicology and Environmental Safety 68, 335-342

Klenell, M., Morita, S., Tiemblo-Olmo, M., Mühlenbock, P, Karpinski, S. and Karpinski, B. 2005. Involvement of the chloroplast signal recognition particle cpSRP43 in acclimation to photooxidative stress in Arabidopsis. Plant Cell Physiology 46, 118-129.

Körner, S. 2001. Development of submerged macrophytes in shallow lake Müggelseee (Berlin, Germany) before and after its switch to the phytoplankton-dominated state. Arch. Hydrobiol. 152, 395-409.

Körner, S. 2002. Loss of submerged macrophytes in shallow lakes in North-eastern Germany. International Review of Hydrobiolia 87, 375-384.

Kobbia, I.A., Battah, M.G., Shabana, E.F., and Eladel, H.M. 2001. Chlorophyll a fluorescence and photosynthetic activity as tools for the evaluation of simazine toxicity to *Protosiphon botryoides* and *Anabaena variabilis*. Ecotoxicology and Environmental Safety 49, 101-105.

Kwak, S.S., Kim, S.K., Park, I.H. and Liu, J.R. 1996. Enhancement of peroxidase activity by stress-related chemicals in sweet potato. Phytochemistry 43, 565-568.

Laing, G.D., Ryckegem, G.V., Tack, F.M.G., Verloo, M.G. 2006. Metal accumulation in intertidal litter through decomposing leaf blades, sheaths and stems of *Phragmites australis*. Chemosphere 63, 1815-1823.

Lara-Nunez, A., Romero-Romero, T., Ventura, J.L., Blancas, V., Anaya, A.L. and Cruz-Ortega, R. 2006. Allelochemical stress causes inhibition of growth and oxidative damage in *Lycopersicon esculentum* Mill. Plant Cell and Environment 29, 2009-2016.

Leenheer, J.A. and Croue, J.-P. 2003. Characterising aquatic dissolved organic matter. Environmental Science and Technology A-Pages 37, 18A-26A.

Li, M., Zhang, L.J., Tao, L., Li, W. 2008. Ecophysiological responses of *Jussiaea rapens* to cadmium exposure. Aquatic Botany 88, 347-352.

Lichtenthaler, H.K. 1996. Vegetation stress: an introduction to the stress concept in plants. Journal of Plant Physiology 148, 4-14.

Madamanchi, N., Alscher, R., Hatzios, K., and Cramer, C. 1994. "Acquired resistance to herbicides in Pea cultivars by exposure to sulphur dioxide." Pesticide Biochemistry and Physiology 48, 31-40.

Madsen, T.V., Maberly, S.C. and Bowes, G. 1996. Photosynthetic acclimation of submersed angiosperms to CO_2 and HCO_3^- Aquatic Botany 53, 15-30.

Madsen, T.V. and Brix, H. 1997. Growth, photosynthesis and acclimation by two submerged macrophytes in relation to temperature. Oecologia 110, 320-327.

Matsunaga, K., Kawaguchi, T., Suzuki, Y. and Nigi, G. 1999. The role of terrestrial humic substances on the shift of Kelp community to crustose coralline algae community of the southern Hokkaido Island in the Japan sea. J. Exp. Mar. Biol. Ecol. 241, 193-205.

Maxwell, C., Griffiths, H., Young, A.J., 1994. Photosynthetic acclimation to light regime and water stress by the C3-CAM epiphyte Guzmania monostachia: Gas-exchange characteristics, photochemical efficiency and the xanthophyll cycle. Functional Ecology 8, 746-754.

May, M.J. and Leaver, C.J. 1993. Oxidative stimulation of glutathione synthesis in *Arabidopsis thaliana* suspension cultures. Plant Physiology 103, 621-627.

Menone, M.L. Pflugmacher, S., 2005. Effects of 3-chlorobiphenyl on photosynthetic oxygen production, glutathione content and detoxication enzymes in the aquatic macrophyte *Ceratophyllum demersum*. Chemosphere 60, 79-84.

Meyer, J.L., Wallace, J.B., Eggert, S.L., 1998. Leaf litter as a source of dissolved organic carbon in streams. Ecosystems 1, 240-249.

Mittova, M., Theodoulou, F.L., Kiddle, G., Volokita, M., Tal, M., Foyer, C.H., Guy, M. 2004. Comparison of mitochondrial ascorbate peroxidase in the cultivated tomato, *Lycopersicum esculentum*, and its wild, salt-tolerant relative, *L. pennellii* – a role for matrix isoforms in protection against oxidative damage. Plant Cell and Environment 27, 237-250.

Montoya-García L, Muñoz-Ocotero, V., Aguilar, R. and de Jiménez, E.S. 2002. Regulation of acidic ribosomal protein expression and phosphorylation in maize. Biochemistry 41, 10166-10172

Mullineaux, P., Ball, L., Escobar, C., Karpinska, B., Greissen, G. and Karpinski, S. 2000. Are diverse signalling pathways integrated in the regulation of Arabidoposis antioxidant defence gene expression in response to excess excitation energy? Philosophical Transactions of the Royal Society of London B Biological Sciences 355, 1531-1540.

Nardi, S., Pizzeghello, D., Muscolo, A. and Vianello, A. 2002. Physiological effects of humic substances on higher plants. Soil Biol. Biochem. 34, 1527-1536.

Nimptsch, J. and Pflugmacher, S. 2005. Substrate specificities of cytosolic glutathione-S transferases in five different species of the aquatic macrophyte *Myriophyllum. Journal of Applied Botany and Food Quality-Angewandte Botanik* 79, 94-99.

Nimptsch, J. and Pflugmacher, S. 2008. Decomposing leaf litter: The effect of allochthonous degradation products on the antioxidant fitness and photosynthesis of *Vesicularia dubyana*. Ecotoxicology and Environmental Safety 69, 541-5.

Noctor, G., Arisi, A.M., Jouanin, L., Kunert, K.J., Rennenberg, H. and Foyer, C.H. 1998. Glutathione: biosynthesis, metabolism and relationship to stress tolerance explored in transformed plants. J. Exp. Bot. 49, 623-647.

Noctor, G., Gomez, L., Vanacker, H., Foyer, C.H. 2002. Interactions between biosynthesis, compartmentation and transport in the control of glutathione homeostais and signalling. Journal of Experimental Botany 53, 1283- 1304.

Noctor, G. 2006. Metabolic signaling in defence and stress: the central roles of soluble redox couples. Plant Cell and Environment 29, 409-425.

Oberg, L.G., Glas, B., Swanson, S.E., Rappe, C. and Paul, K.G. 1990. Peroxidas-catalyzed oxidation of chlorophenols to polychlorinated dibenzo-p-dioxins and dibenzofurans. Archives of Environmental Contamination Toxicology 19, 930-938.

OECD 2006: Lemna sp. Growth inhibition test. Guidelines (221) for the testing of chemicals. Organisation for Economic Cooperation and Development, Berlin.

Oettmeier, W., Masson, K. and Donner, A. 1988. Anthraquinone inhibitors of photosystem II electron transport. *FEBS Letters* 231, 259-262.

Olesen, B. and Madsen, T.V. 2000. Growth and physiological acclimation to temperature and inorganic carbon availability to two submerged aquatic macrophyte species, *Callitriche cophocarpa* and *Elodea Canadensis.* Functional Ecology 14, 252-260.

Penuelas, J., Llusia, J., Asensio, D. and Munne-Bosch, S. 2005. Linking isoprene with plant thermotolerance, antioxidants and monoterpene emissions. Plant Cell and Environment 28, 278-286.

Pfannschmidt, T. (2003) Chloroplast redox signals: how photosynthesis controls its own genes. Trends in Plant Sciences 8, 33-41.

Pflugmacher, S. and Steinberg, C.E.W., 1997. Activity of phase I and phase II detoxication enzymes in aquatic macrophytes. Journal of Applied Botany 71, 144-146.

Pflugmacher, S., Spangenberg, M., Steinberg, C.E.W. 1999. Dissolved organic matter (DOM) and effects on the aquatic macrophyte *Ceratophyllum demersum* in relation to photosynthesis, pigment pattern and activity of detoxication enzymes. Journal of Applied Botany 73, 184-190.

Pflugmacher, S., Pietsch, C., Reiger, W., Paul, A., Preuer, T., Zwirnmann, E., Steinberg C.E.W., 2003. Humic substances and their direct effects on the physiology of aquatic plants. In: Ghabbour, E.A., Davies, G. (Eds.), *Humic substances: nature's most versatile materials*. Taylor and Francis, New York.

Pflugmacher, S., Pietsch, C., Rieger, W. and Steinberg, C.E.W. 2006. Dissloved natural organic matter impacts photosynthetic oxygen production and electron transport in coontail *Ceratophyllum demersum*. Sci. Tot. Environ. 357, 169-175.

Price, A.H., Atherton, N.M., Hendry, G.A.F., 1989. Plants under drought-stress generate activated oxygen. Free Radical Res. Commun.8, 61-66.

Razinger, J., Dermastia, M., Drinovec, L., Drobne, D., Zrimec, A. and Koce, J.D. 2007. Antioxidative responses of Duckweed (*Lemna minor* L.) to short-term copper exposure. Environmental Science and Pollution Research 14, 194-201.

Rennenberg, H. 1982. Glutathione metabolism and possible biological roles in higher plants. Phytochemistry 21, 2771-2781.

Rice-Evans, C.A., Miller, N.J., Bolwell, P.G., Bramley, P.M. and Pridham, J.B. 1995. The relative antioxidant activities of plant-derived polyphenolic flavonoids. Free Radical Res. 22, 375-383.

Ridge, I., Walters, J. and Street, M. 1999. Algal growth control by terrestrial leaf litter: A realistic tool? Hydrobiologia 396, 173-180.

Rijstenbil, J. W., Haritonidis, S., Malea, P., Seferlis, M. and Wijnholds, J.A. 1998. Thiol pools and glutathione redox ratios as possible indicators of copper toxicity in the green macroalgae *Enteromorpha* spp. from the Scheldt Estuary (SW Netherlands, Belgium) and Thermaikos Gulf (Greece, N Aegean Sea). Hydrobiologia 385, 171-181.

Robinson, C.T. and Gessner, M.O. 2000. Nutrient addition accelerates leaf breakdown in an alpine springbrook. Oecologia 122, 258-263.

Romero-Puertas, M. C., Corpas, F.C., Sandalio, L.M., Leterrier, M., Rodriguez-Serrano, M., del Rio, L.A. and Palma, J.M. 2006. Glutathione reductase from pea leaves: response to abiotic stress and characterization of the peroxisomal isozyme. New Phytology 170, 43-52.

Sachse, A., Babenzien, D., Ginzel, G., Gelbrecht, J., Steinberg, C.E.W., 2001. Characterization of dissolved organic carbon (DOC) in a dystrophic lake and an adjacent fen. Biogeochemistry 54, 279–296.

Salminen, J., Roslin, T., Karonen, M., Sinkkonen, J., Pihlaja, K., Pulkkinen, P., 2004. Seasonal variation in the content of hydrolizable tannins, flavonoid glycosides and proanthocyanidins in oak leaves. Journal of Chemical Ecology 30, 1693-1711.

Sandermann, H.(1994) Higher plant metabolism of xenobiotics: the ‘green liver’ concept. Pharmacogenetics 4, 225-241.

Sarvala, J., Ilmavirta, V., Paasivirta, L. and Salonen, K. 1981. The ecosystem of the oligotrophic lake Pääjärvi 3. Secondary production and an ecological budget of the lake. Verhandlung der Internationalen Vereinigung für Limnologie 21, 422-427.

Scalbert, A., Monties, B., Favre, J.M., 1988. Polyphenols of *Quercus robur*: Adult tree and in vitro grown calli and shoots. Phytochemistry 27,: 3483-3488.

Scandalios, J.G. 2005. Oxidative stress: molecular perception and transduction of signals trigerring antioxidant gene defenses. Brazilian Journal of Medical and Biological Research 38, 995-1014.

Scheibe, R., Backhausen, J.E., Emmerlich, V. and Holtgrefe, S. 2005. Strategies to maintain redox homeostasis during photosynthesis under changing conditions. Journal of Experimental Botany 56, 1481-1489.

Schneider, C., Tallman, K.A., Porter, N.A. and Brash, A.R. 2001. Two distinct pathways of formation of 4-hydroxynonenal. Mechanisms of non-enzymatic transformation of the 9- and 13-hydroperoxides of linoleic acid to 4-hydroxyalkenals. Journal of Biological Chemistry 276, 20831-20838.

Schröder, P. 1997. Fate of glutathione S-conjugates in plants: cleavage of the glutathione moiety. In: Hatzios, K.K. (ed.) Regulation of enzymatic systems detoxifying xenobiotics in plants. NATO ASI Series Vol. 37, Kluwer NL, pp. 233-244.

Sersen, F., Kralova, K. and Bumbalova, A. 1998. Action of mercury on the photosynthetic apparatus of spinach chloroplasts. Photosynthetica 35, 551-559.

Soleto D.S. and Khanna-Choppra, R. 2006. Drought acclimation confers oxidative stress tolerance by inducing co-ordinated antioxidant defense at cellular and subcellular level in leaves of wheat seedlings. Physiologia Plantarum 127, 494-506.

Streb, P. and Feierabend, J. 1996. Oxidative stress responses accompanying photoinactivation of catalase in NaCl-treated rye leaves. Botanica Acta 109, 125-132.

Tausz, M., Sircelj, H. and Grill, D. 2004. The glutathione system as a stress marker in plant ecophysiology: is a stress-response concept valid? Journal of Experimental Botany 55, 1955- 1962.

Teisseire, H. and Guy, V. 2000. Copper-induced changes in antioxidant enzymes activities in fronds of duckweed (Lemna minor). Plant Science 153, 65-72.

Thurman, E.M., 1985. Organic geochemistry of natural waters, Kluwer, Dordrecht, pp. 497.

Tiegs, S.D., Peter, F.D., Robinson, C.T., Uehlinger, U. and Gessner, M.O. 2008. Leaf decomposition and invertebrate colonization responses to manipulated litter quantity in streams. Journal of North American Benthological Society 27, 321-331.

Tkalec, M., Malaric, K. and Pevalek-Kozlina, B. 2007. Exposure to radiofrequency radiation induces oxidative stress in duckweed *Lemna* minor L. Science of the Total Environment 388, 78-89.

Turgut, C. and Fomin, A. 2002. Sensitivity of the rooted macrophyte *Myriophyllum aquaticum* (Vell.) Verdcourt to seventeen pesticides determined on the basis of EC_{50}. Bulletin of Environmental Contamination and Toxicology 69, 601-608.

Vanacker, H., Carver, T.L.W. and Foyer, C.H. 2000. Early H_2O_2 accumulation in mesophyll cells leads to induction of glutathione during the hyper-sensitive response in the Barley-Powdery Mildew interaction. Plant Physiology 123, 1289-1300.

V.-Balogh, K., Presing, M., Vörös, L. and Toth, M. 2006. A study of the decomposition of reed (Phragmites australis) as possible source of aquatic humic substances by measuring the natural abundance of stable carbon isotopes. International Review of Hydrobiologia 91, 15-28.

Viet, U. and Moser, H. 2003. Ecotoxicological test assays with macrophytes. Fresenius Environmental Bulletin 12, 545-549.

Vranova, E., Atichartpogkul, S., Villarroel, R., Van Montagu, M., Inze, D. and Van Camp, W. 2002. Comprehensive analysis of gene expression in *Nicotiana tabacum* leaves acclimated to oxidative stress. Proc. Nat. Acad. Sci., USA 99, 10870-10875.

Wang, W.H., Bray, C.M. and Jones, M.N. 1999. The fate of ^{14}C-labelled humic substances in rice cells in cultures. Journal of Plant Physiology 154, 203-211.

Ward, G.M., Ward, A.K., Dahm, C.N., Aumen, N.G. 1994. Origin and formation of organic and inorganic particles in aquatic systems. In: Worton, R.S. (ed.) The biology of particles in aquatic systems. 2nd ed. CRC press Inc. pp. 45-74. (ISBN 0873719050)

Weir, T.L., Park, S. and Vivanco, J.M., 2004 Biochemical and physiological mechanisms mediated by allelochemicals. Current Opinion in Plant Biology 7, 472-479.

Wetzel, R.G., 2001. Limnology: lake and river ecosystems. Third ed. Academic Press, San Diego. pp. 527-664.

Wetzel, R.G., 1995. Death, detritus and energy flow in aquatic ecosystems. Freshwater Biology 33, 83-89.

Willekens, H., Villaroel, R., Van Montagu M., Inze, D., and Van Camp, W. 1994. Molecular identification of catalases from *Nicotiana plumbaginifolia* (L). FEBS Lett. 352, 79-83.

Xiang, C. and Oliver, D.J. 1998. Glutathione metabolic genes coordinatedly respond to heavy metals and jasmonic acid in Arabidopsis. Plant Cell 10, 1539-1550.

Yang, Y., Sharma, R., Sharma, A., Awasthi, S. and Awasthi, Y.C. 2003. Lipid peroxidation and cell cycle signalling: 4-hydroxynonenal, a key molecule in stress mediated signaling. Review. Acta Biochimica Polonica 50, 319-336.

Zar, J.H. 1996. Biostatistical analysis. Third ed. Pentrice-Hall, Inc. New Jersey 07458.

Zhu, Y.L., Philon-Smits, E.A,H., Tarun, A.S., Weber, S.U., Jouanin, L. and Terry, N. 1999. Cadmium tolerance and accumulation in Indian mustard is enhanced by overexpressing gamma-glutamylcysteine synthetase. Plant Physiology 121, 1169-1178.

List of abbreviations

ANOVA	Analysis of variance
CAT	Catalase
CDNB	1-chloro-2,4-dinitrobenzene
cDNA	complementary DNA
cGST	Cytosolic GST
DMF	Dimethylformamide
DMSO	Dimethylsulphoxide
DNA	Deoxyribonucleic acid
DNAase	Deoxyribonuclease
dNTP	Deoxyribonucleotide triphosphate
DOC	Dissolve organic carbon
DTE	Dithioerythritol
DTNB	5,5-dithio-2-nitrobenzoic acid
DW	Dry weight
EDTA	Ethylenediaminetetracetic acid
FN	Frond number
FW	Fresh weight
GAPDH	Glyceraldehyde-3-phosphate dehydrogenase
GPx	Glutathione peroxidase
GR	Glutathione reductase
GST	Glutathione S-transferase
GSH	Reduced glutathione
GSSG	Glutathione disulphide
HS	Humic substances
HMWS	High molecular weight substances
HNE	Hydroxynonenal
L	Litre
LMWS	Low molecular weight substances
LOEC	Lowest observed effect concentration
LPO	Lipid peroxidaton
MDA	Malondialdehyde
mg	Milligram
mGST	Membrane bound GST
mL	Millilitre
mRNA	Messenger RNA
NADPH	reduced nicotinamide adenine dinulcleotide phosphate
NAP-5	Sephadex G-25 Column
nkat	nano katal
NOEC	No observed effect concentration
NOM	Natural organic matter

OECD	Organisation for economic co-operation and development
PC	Plastocyanin
PCR	Polymerase chain reaction
POD	Peroxidase
PQ	Plastoquinone
PS	Photosystem
RNA	Ribonucleic acid
ROS	Reactive oxygen species
RT	Reverse transcriptase
SOD	Superoxide dismutase

Declaration

I hereby declare that this dissertation was independently written without any external assistance and that no sources other than those cited were used. Furthermore, I declare that this work is not under consideration for submission to another examining institution.

Berlin,

Sheku Kamara

Publications and Conference participation

Peer-reviewed journal publications

Kamara, S., Pflugmacher, S. 2007. *Phragmites australis* and *Quercus robur* leaf extracts affect antioxidative system and photosynthesis of *Ceratophyllum demersum*. Ecotoxicology and Environmental Safety 67, 240-246.

Kamara, S., Pflugmacher, S. 2007. Acclimation of *Ceratophyllum demersum* to stress imposed by *Phragmites australis* and *Quercus robur* leaf extracts. Ecotoxicology and Environmental Safety 68, 335-342 (**Highlighted Article**)

Kamara, S., Pflugmacher, S. 2008. Effects of leaf extracts on glutathione reductase expression, hydrogen peroxide and glutathione contents in the aquatic macrophyte *Ceratophyllum demersum*. Aquatic Sciences 70, 204-211.

Kamara, S., Pflugmacher, S. (submitted). Growth and antioxidative mechanisms in *Lemna minor* due to leaf litter-derived dissolved organic carbon. Chemico-Biological Interactions (*in review*)

International Conference Presentations:

Platform presentations

Kamara, S., Pflugmacher, S. Effects of leaf extracts from *Phragmites australis* and *Quercus robur* on the antioxidative system of the aquatic macrophyte *Ceratophyllum demersum*, Society of Environmental Toxicology and Chemistry (SETAC) Europe 16th Annual Meeting, The Hague, (Netherlands) 07 - 11 May, 2006.

Kamara, S., Pflugmacher, S. Effect of leaf extracts on the glutathione reductase gene expression and the antioxidative systems of the aquatic macrophyte *Ceratophyllum demersum*. Society of Environmental Toxicology and Chemistry (SETAC) Africa/ANCAP International conference, Arusha, Tanzania, 16 – 20 October, 2006.

Kamara, S., Pflugmacher, S. Phytotoxicity of leaf extracts on the aquatic macrophyte *Lemna minor*. Society of Environmental Toxicology and Chemistry (SETAC) North America 28th Annual Meeting in Milwaukee (USA) 11 – 15 November 2007.

Kamara, S., Pflugmacher, S. (To be presented). Oxidative stress response of the aquatic macrophyte *Lemna minor* due to leaf leachates exposure. Society of Environmental Toxicology and Chemistry (SETAC) World Congress in Sydney, Australia, 3-7 August, 2008.

Poster presentations

Kamara, S., Pflugmacher, S. Acclimation of *Ceratophyllum demersum* to physiological stress imposed by reed (*Phragmites australis*) and oak (*Quercus robur)* leaf extracts SETAC Europe 17th Annual Meeting, Porto, Portugal. May 20-24, 2007.

Kamara, S., Pflugmacher, S. Changes in antioxidative gene regulation in *Ceratophyllum demersum* due to the influence of leaf leachates. SETAC Europe 17th Annual Meeting, Porto, Portugal. May 20-24, 2007.

Acknowledgements

I am profoundly grateful to PD Dr. Stephan Pflugmacher for guiding me through this study. Despite his busy schedule, he always spared some of his valuable time to hold critical discussions with me regarding my research.

My gratitude also goes to Prof. Dr. Claudia Wiegand for her complementary academic advice and support.

Alongside academic challenges, one is also occasionally confronted with private issues. In this regard, I am highly indebted to Prof. Dr. Werner Kloas for patiently seeking solutions to my private problems.

Dr. Jorge Nimptsch, I appreciate your role as my immediate academic mentor.

My thanks and appreciation to all past and present members (Maria Vassilakaki, Anja Peuthert, Reda Grigutyte, Valeska Contardo, Viola Viehmann, Bettina Hübner, Christoph Bahrdt, Rafael Ortiz, Katrin Hinrichs) of the research group, biochemical regulation, for providing various kinds of assistance and an enabling environment to endure the difficult and challenging times in the lab.

Secretariat of IGB, especially Kathrin Lehmann, is gratefully acknowledged for arganisational and administrative assistance. Members of the Chemical Laboratory of the IGB, especially Dr. Elke Zwirnmann, are highly appreciated for assisting with chemical analyses of leaf litter extracts.

Members of the Sierra Leone community in Berlin, notably Dr. Karifala Dumbuya and Madam Adama Thorlie, are gratefully acknowledged for being there for me at all times. You made it easy for me to acclimatize in Berlin and I thank you for that.

Behind every successful man is a woman. Thanks to my wife, Nancy Kamara, who always asked: When are you coming back today? And got the response: I don't know. She sacrificed everything to come and provide support during my study in Germany. I am grateful for your patience. I thank my lovely daughters, Shekunatu and Mariam, who accepted less than the time and attention they deserved from me.

It is impossible to mention everyone. Therefore, my sincerest thanks go to all those who helped me in various ways but whose names are not listed here.

Above all, I thank God for everything.

Curriculum Vitae

1. Personal details

First name Sheku
Last name Kamara
Nationality Sierra Leonean
Date of birth 20.02.70
Place of birth Timbo Village (Tonko Limba Chiefdom)
Email: kamara@igb-berlin.de OR mamadie2001@yahoo.com

2. Academic details

1988 - 1990 University-entrance GCE 'O' Level (First Division), West African Examinations Council
1990 - 1996 BSc (Hons. Second Class, First Division) Botany, Fourah Bay College, University of Sierra Leone
Oct. 2001 - Jun. 2003 MSc Environmental Science and Technology, International Institute for Hydraulic and Environmental Engineering (IHE), The Netherlands.
Oct. 2002 - Mar. 2003 Research stay at EAWAG, Switzerland, Preparation of MSc thesis on 'Effects of temperature on decomposition of *Phragmites australis* and macroinvertebrate assemblage in the littoral zone of lake Hallwil, Switzerland'.
Since April 2005 PhD student at the Humboldt University in Berlin; Research stay at the Leibniz Institute of Freshwater Ecology and Inland Fisheries (IGB) Berlin working on the topic 'Physiological responses of aquatic macrophytes to natural organic matter: potential for structuring aquatic ecosystems'.

3. Employment

1996 – 2001 Temporary Teaching Assistant (TTA), Department of Biological Sciences, Fourah Bay College, University of Sierra Leone, Freetown, Sierra Leone.
Mar. – Oct. 2004 Assistant Lecturer, Fourah Bay College, University of Sierra Leone, Freetown, Sierra Leone.
Jun. 2004 – Oct. 2004 Expert UN Climate Change Project on National Inventory of Green House Gas Emissions: In charge of Land Use, Land Use Change and Forestry (LULUCF).

4. Membership in Professional Organisation

Member: Society of Environmental Toxicology and Chemistry (SETAC) - Africa/Europe branch

5. Scholarships

1991-1996 Sierra Leone Government Grant-in-Aid (SLG) – B.Sc. study
2001-2003 Netherlands Fellowship Programme (NFP) – M.Sc. study
2004-2008 German Academic Exchange Service (DAAD) – Ph.D. study